Applied Intelligence for Industry 4.0

We all are aware that artificial intelligence (AI) has brought a change in our lives, driven by a new form of interaction between man and machine. We are in the era of the fourth Industrial Revolution (IR) where AI plays vital roles in human development by enabling extraordinary technological advances and making fundamental changes to the way we live, work and relate to one another. It is an opportunity to help everyone, including leaders, policymakers and people from all income groups and nations, to harness converging technologies in order to create an inclusive, human-centered future. We need to prepare our graduates as well as researchers to conduct their research with 4.0 IR-related technologies. We need to develop policies and implement them to focus on the components of 4.0 IR for sustainable developments. *Applied Intelligence for Industry 4.0* will cover cutting edge topics in the fields of AI and Industry 4.0. It will appeal to beginners and advanced researchers in computer science, information sciences, engineering and robotics.

Key Features

- Discusses advanced data mining, feature extraction and classification algorithms for disease detection, cyber security detection and prevention, soil quality assessment and other industrial applications
- Includes the parameter optimization and explainablity of intelligent approaches for business applications
- Presents context-aware smart insights and energy efficient and smart computing for next-generation smart industry

Applied Intelligence for Industry 4.0

Edited by
Nazmul Siddique
Mohammad Shamsul Arefin
M Shamim Kaiser
ASM Kayes

CRC Press is an imprint of the
Taylor & Francis Group, an **informa** business
A CHAPMAN & HALL BOOK

First edition published 2023
by CRC Press
6000 Broken Sound Parkway NW, Suite 300, Boca Raton, FL 33487-2742

and by CRC Press
4 Park Square, Milton Park, Abingdon, Oxon, OX14 4RN

CRC Press is an imprint of Taylor & Francis Group, LLC

Library of Congress Cataloging-in-Publication Data

Names: Siddique, N. H., editor.
Title: Applied intelligence for industry 4.0 / edited by Nazmul Siddique, Mohammad Shamsul, M. Shamim Kaiser, and A.S.M. Kayes.
Description: First edition. | Boca Raton, FL : CRC Press, 2023. | Includes bibliographical references and index.
Identifiers: LCCN 2022029013 (print) | LCCN 2022029014 (ebook) | ISBN 9781032164151 (hbk) | ISBN 9781032187556 (pbk) | ISBN 9781003256083 (ebk)
Subjects: LCSH: Artificial intelligence--Industrial applications.
Classification: LCC Q335 .A473 2023 (print) | LCC Q335 (ebook) | DDC 006.3--dc23/eng20221031
LC record available at https://lccn.loc.gov/2022029013
LC ebook record available at https://lccn.loc.gov/2022029014

ISBN: 978-1-032-16415-1 (hbk)
ISBN: 978-1-032-18755-6 (pbk)
ISBN: 978-1-003-25608-3 (ebk)

DOI: 10.1201/9781003256083

Typeset in Latin Modern font
by KnowledgeWorks Global Ltd.

Publisher's note: This book has been prepared from camera-ready copy provided by the authors.

This book is dedicated to all the researchers, professionals and students who are continuously trying their best for the advancement of technologies for the betterment of humanity.

Contents

Preface

The digital enterprise, which is already a reality, is growing increasingly prevalent in industry. Data is generated, processed, and evaluated on a continuous basis. The large amounts of data created in manufacturing environments serve as the foundation for the creation of digital representations of entire plants and systems. The use of digital twins has been around for quite some time to structure the planning and design of products and machinery – and even the actual manufacturing operations themselves – and to do so more flexibly and efficiently while producing high-quality, customized products faster and at a lower cost. An effective automation and AI solution can transform industry. The automated machines and processes can collect insights from these massive amounts of data, and then use those insights to improve the efficiency of the operations of these machines.

Artificial intelligence, sensors, and digital platforms have created unprecedented opportunities for rapid learning. These technologies are overturning industrial markets and compelling businesses to rethink how they accomplish traditional tasks such as optimization and control, condition monitoring, process advice and troubleshooting, reconfigurable production, and intelligent dashboards.

Industry 4.0 is already a reality for a significant number of businesses worldwide. Nonetheless, it is worth noting that the modifications necessary to fully participate in this industrial revolution and reap its rewards will not occur suddenly. As with digital transformation and automation, this is a continuous improvement process. As can be seen, AI is a critical component of this transformation.

This book presents a variety of artificial intelligence-based models for sentiment analysis, disease detection and prediction, decision support system design, forecasting model development, soil quality assessment, COVID-19 screening, and ensuring countermeasures against various forms of attacks. The book can be used as a reference for advanced undergraduate or graduate students studying computer science or related subjects such as software engineering, information and communication engineering, security, industrial automation, etc. It will appeal more to those with a technical bent; sections are technically demanding and emphasize learning by experience: designing, constructing, and implementing systems.

Contributors

Abdullah Bin Shams
Dept. of ECE, University of Toronto,
Canada

Abu Zahid Bin Aziz
Rajshahi University of Engineering & Technology,
Rajshahi, Bangladesh

Abdul Kadar Muhammad Masum
Department of Computer Science and Engineering, International Islamic University Chittagong,
Chattogram, Bangladesh

A.N.M. Rezaul Karim
Department of Computer Science and Engineering, International Islamic University Chittagong,
Chattogram, Bangladesh

Arifa Islam Champa
Bangladesh Army International University of Science and Technology,
Comilla, Bangladesh

Anik Sen
Department of Computer Science and Engineering, Premier University,
Chattogram, Bangladesh

Annesha Das
Department of Computer Science and Engineering, Chittagong University of Engineering and Technology,
Chattogram, Bangladesh

Ashim Dey
Department of Computer Science and Engineering, Chittagong University of Engineering and Technology,
Chattogram, Bangladesh

Diti Roy
Institute of Information and Communication Technology Khulna University of Engineering and Technology (KUET),
Khulna, Bangladesh

Dipon Talukder
Chittagong University of Engineering and Technology,
Chattogram, Bangladesh

Hasan Murad
Dept. of CSE, University of Asia Pacific,
Dhaka, Bangladesh

Hasin Rehana
Rajshahi University of Engineering and Technology,
Rajshahi, Bangladesh

Himadri Sikder Badhon
North Western University,
Khulna, Bangladesh

Iqbal H. Sarker
Department of Computer Science and Engineering, Chittagong University of Engineering and Technology,
Chattogram, Bangladesh

Kazi Mowdud Ahmed
Islamic University,
Kushtia, Bangladesh

Kaushik Deb
Chittagong University of Engineering & Technology,
Chattogram, Bangladesh

Khandaker Tayef Shahriar
Department of Computer Science & Engineering, Chittagong University of Engineering & Technology,
Chittagong, Bangladesh

Khaleque Md. Aashiq Kamal
Department of Computer Science & Engineering, Premier University,
Chattogram, Bangladesh

Kamruzzaman Chowdhury
Department of Computer Science & Engineering, East Delta University,
Chattogram, Bangladesh

Khaleque Md. Aashiq Kamal
King Fahd University of Petroleum and Minerals,
Dhahran, KSA

Laboni Akter
Dept. of BME, Khulna University of Engineering & Technology,
Khulna, Bangladesh

M. Raihan
North Western University,
Khulna, Bangladesh

Md Shamim Hossain
Islamic University,
Kushtia, Bangladesh

Md. Fazle Rabbi
Bangladesh Army International University of Science and Technology
Comilla, Bangladesh

Md. Golam Hafez
Department of Mathematics, Chittagong University of Engineering and Technology,
Chattogram, Bangladesh

Md. Ali Akbar
Department of Computer Science & Engineering, East Delta University,
Chattogram, Bangladesh

Md. Asif Zaman
Rajshahi University of Engineering and Technology,
Rajshahi, Bangladesh

Maimuna Manita Hoque
Department of Computer Science & Engineering, East Delta University,
Chattogram, Bangladesh

Mohammed Nazim Uddin
School of Science, Engineering & Technology, East Delta University,
Chattogram, Bangladesh

Md. Neamul Haque
Premier University,
Chattogram, Bangladesh

Md. Mohsin Sarker Raihan
Dept. of BME, Khulna University of Engineering & Technology,
Khulna 9203, Bangladesh

Md. Mohi Uddin Khan
Dept. of EEE, Islamic University of Technology, Boardbazar,
Gazipur, Bangladesh

Md. Rifat Hossain
Bangladesh Army International University of Science and Technology,
Comilla, Bangladesh

Mohammad Ali Moni
WHO Collaborating Centre on eHealth, UNSW Digital Health, Faculty of Medicine, University of New South Wales, Sydney,
NSW, Australia

Moinul Islam Sayed
Patuakhali Science and Technology University,
Patuakhali, Bangladesh

Muhammad Nazrul Islam
Department of Computer Science and Engineering, Military Institute of Science and Technology,
Dhaka, Bangladesh

Mohammed Nazim Uddin
School of Science, Engineering & Technology, East Delta University,
Chattogram, Bangladesh

Md. Ashiq Mahmood
Institute of Information and Communication Technology Khulna University of Engineering & Technology,
Khulna, Bangladesh

Md. Mokammel Haque
Chittagong University of Engineering and Technology,
Chittagong, Bangladesh

Md. Musfique Anwar
Department of Computer Science and Engineering, Jahangirnagar University,
Dhaka, Bangladesh

Md. Al Mamun
Rajshahi University of Engineering & Technology,
Rajshahi, Bangladesh

Md. Palash Uddin
Deakin University,
Australia

Md Rahat Hossain
Central Queensland University, North Rockhampton,
QLD, Australia

Md Khairul Islam
Islamic University,
Kushtia, Bangladesh

Md Mahbubur Rahman
Islamic University,
Kushtia, Bangladesh

Nabila Nawal
Department of Computer Science and Engineering, Chittagong University of Engineering and Technology,
Chittagong, Bangladesh

Nazma Akther
Premier University,
Chattogram, Bangladesh

Nishat Tasnim Tonni
North Western University,
Khulna, Bangladesh

Nuren Nafisa
Department of Computer Science and Engineering, Chittagong University of Technology and Engineering,
Chattogram, Bangladesh

Nusrat Jahan Euna
Department of Computer Science & Engineering, Chittagong University of Engineering & Technology,
Chattogram, Bangladesh

Promila Ghosh
North Western University,
Khulna, Bangladesh

Rashik Rahman
Dept. of CSE, University of Asia Pacific,
Dhaka, Bangladesh

Rejwana Islam
Global University Bangladesh,
Barishal, Bangladesh

Saadman Sakib
Chittagong University of Engineering & Technology,
Chattogram, Bangladesh

Sabrina Jahan Maisha
Department of Computer Science and Engineering, BGC Trust University Bangladesh,
Chattogram, Bangladesh

Sharmin Akter
Department of Computer Science and Engineering, Chittagong University of Engineering and Technology,
Chittagong, Bangladesh

Sharmin Akter
Department of Computer Science and Engineering, Chittagong University of Engineering and Technology,
Chittagong, Bangladesh

Md Sipon Miah
Islamic University,
Kushtia, Bangladesh

S. M. Mahedy Hasan
Rajshahi University of Engineering and Technology,
Rajshahi, Bangladesh

Sayed Allamah Iqbal
Department of Electrical and Electronic Engineering, International Islamic University Chittagong,
Chattogram, Bangladesh

Sayed Asaduzzaman
Rangamati Science and Technology University,
Rangamati, Bangladesh

Sajal Saha
Patuakhali Science and Technology University,
Patuakhali, Bangladesh

Syed Mohd. Farhan
Department of Computer Science & Engineering, East Delta University,
Chattogram, Bangladesh

Syed Md. Minhaz Hossain
Department of Computer Science & Engineering, Premier University,
Chattogram, Bangladesh

Tamal Joyti Roy
Institute of Information and Communication Technology Khulna University of Engineering & Technology,
Khulna, Bangladesh

Editors

Nazmul Siddique is with the School of Computing, Engineering and Intelligent Systems, Ulster University. He obtained his Dipl.-Ing. degree in Cybernetics from the Dresden University of Technology, Germany, his MSc in Computer Science from Bangladesh University of Engineering and Technology, and his PhD in Intelligent Control from the Department of Automatic Control and Systems Engineering, University of Sheffield, England. His research interests include: cybernetics, computational intelligence, nature-inspired computing, stochastic systems and vehicular communication. He has published over 195 research papers including five books published by John Wiley, Springer and Taylor & Francis. He guest-edited eight special issues of reputed journals on Cybernetic Intelligence, Computational Intelligence, Neural Networks and Robotics. He is on the editorial board of seven international journals including *Nature Scientific Reports*. He is a Fellow of the Higher Education Academy, and a senior member of IEEE. He has been involved in organizing many national and international conferences and co-edited seven conference proceedings.

Mohammad Shamsul Arefin is in leave from Chittagong University of Engineering and Technology (CUET) and currently affiliated with the Department of Computer Science and Engineering (CSE), Daffodil International University, Bangladesh. Earlier he was the Head of the Department of CSE, CUET. Prof. Arefin received his Doctor of Engineering Degree in Information Engineering from Hiroshima University, Japan with support of a scholarship from MEXT, Japan. As a part of his doctoral research, Dr. Arefin was with IBM Yamato Software Laboratory, Japan. His research includes privacy preserving data publishing and mining, distributed and cloud computing, big data management, multilingual data management, semantic web, object oriented system development and IT for agriculture and environment. Dr. Arefin has more than 120 refereed publications in international journals, book series and conference proceedings. He is a senior member of IEEE, Member of ACM, and Fellow of IEB and BCS. Dr. Arefin is the Organizing Chair of BIM 2021; TPC Chair, ECCE 2017; Organizing Co-Chair, ECCE 2019; and Organizing Chair, BDML 2020. Dr. Arefin visited Japan, Indonesia, Malaysia, Bhutan, Singapore, South Korea, Egypt, India, Saudi Arabia and China for different professional and social activities.

Dr. M Shamim Kaiser is currently working as a professor at the Institute of Information Technology of Jahangirnagar University, Savar, Dhaka-1342, Bangladesh. He received his Bachelor's and Master's degrees in Applied Physics Electronics and Communication Engineering from the University of Dhaka, Bangladesh, in 2002 and 2004, respectively, and a PhD degree in Telecommunication Engineering from the Asian Institute of Technology, Thailand, in 2010. His current research interests include data analytics, machine learning, wireless network and signal processing, cognitive radio network, big data and cyber security, and renewable energy. He has authored more than 100 papers in different peer-reviewed journals and conferences. He is Associate Editor of the Journal IEEE Access , Guest Editor of *Brain Informatics and Cognitive Computation.* He is a Life Member of the Bangladesh Electronic Society; Bangladesh Physical Society. He is also a senior member of IEEE, USA, and IEICE, Japan, and an active volunteer of the IEEE Bangladesh Section. He is the founding Chapter Chair of the IEEE Bangladesh Section Computer Society Chapter. He has organized various international conferences such as ICEEICT 2015-2018, IEEE HTC 2017, IEEE ICREST 2018 and BI2020.

ASM Kayes is currently a Discipline Coordinator for the Bachelor of Cybersecurity Program in the Department of Computer Science and Information Technology at La Trobe University. He is also the Coordinator for the Master of Cybersecurity subjects. Recently, he has been elected to the position of La Trobe University Academic Board, 2020-2021. Dr Kayes is a cybersecurity expert with many years of research experience. He is passionate about safeguarding data, systems and people from cyber-attacks. He has broad contemporary knowledge of cybersecurity and has deep research and technical skills in data privacy and security. He is currently supervising many PhD students at La Trobe and Victoria Universities. Dr Kayes has collaborated with many distinguished scientists within Australia and internationally from the USA, UK and so on. His research outputs have been published in top-tier journals and conferences, such as *Future Generation Computer Systems*, *Information Systems*, *The Oxford Computer Journal*, *CAiSE*, *ICSOC*, *WISE*, *IEEE TrustCom*, and so on. He has already been recognized as an established researcher in the areas of data privacy, security and access control. His proposed security and access control models have been cited over 300 times in last two years. He has collaborated with industry partners and has been awarded over 415k AUD funding for cybersecurity, privacy, machine learning and deep learning R&D projects.

CHAPTER 1

Multi-labelled Bengali Public Comments Sentiment Analysis with Bidirectional Recurrent Neural Networks (Bi-RNNs)

Promila Ghosh

North Western University, Khulna, Bangladesh

M. Raihan

North Western University, Khulna, Bangladesh

Nishat Tasnim Tonni

North Western University, Khulna, Bangladesh

Himadri Sikder Badhon

North Western University, Khulna, Bangladesh

Sayed Asaduzzaman

Rangamati Science and Technology University, Rangamati,Bangladesh

Hasin Rehana

Rajshahi University of Engineering and Technology, Rajshahi, Bangladesh

CONTENTS

A SENTIMENT ANALYSIS, is one of the most prominent research topics in today's Natural Language Processing (NLP) field to analyze the statements or opinions of individuals. Individuals' statements can be classified into different classes as positive, slightly negative, strongly negative or neutral. In this approach, multi-classified sentiments have been classified or analyzed using a Deep Learning (DL) algorithm named Bidirectional Recurrent Neural Networks (Bi-RNNs) applied on Bengali text data. To analyze people's comments on social sites or e-commerce sites, sentiment analysis can play a notable role. The selected dataset for this approach has been multi-classified with mainly 7 types of sentiment tags based on polarity to acknowledge the sentiment class with more specification. As one of the types of Recurrent Neural Networks (RNN), Bidirectional Recurrent Neural Networks (Bi-RNNs) operate two RNNs, and the inputs are accessed in both forward and reverse directions. We have adopted multi-class classification with Bidirectional Recurrent Neural Networks (Bi-RNNs) and acquired 88% exactitude.

1.1 INTRODUCTION

Sentiment analysis (SA) is a Natural Language Processing (NLP) approach implemented to regulate text data positively, negatively or neutraly. SA is often performed on text data to support commercial brand and product sentiment in customer commentary and recognize customer necessity. A sentiment analysis system for text analysis combines NLP and Machine Learning (ML) methods which are applied to allow loaded sentiment rates to the entities, tags and types within a sentence.

SA assists data interpreters with complex attempts to estimate public evaluation, counsellor label and merchandise credit, and assume customer activities. In joining, data analytics firms often integrate third-party SA APIs into their customer activity administration, social media observation, or human resources analytics policies, to pass valuable insights to their clients [1]. Nowadays people use social media, blog sites, news sites or others to express as well as share their statements or opinions in regular life. It's become a remarkable change in our society through being a two-edged aspect of our society. This has also become an amazing technology for an e-commerce site. Customers find good quality products with price consistency and also can choose their desirable products by others' feedback. The merchants perceive the potential customers and their preference and get their feedback [1].

The spread of rumours promotes social chaos. Often people become offended due to the rumours and commit unethical works. People sometimes share false news and try to harass others through comments, and social degradation is increasing, which has affected mostly the young generation. Moreover, some corrupted e-commerce sites cheat people with their fake products or take money without giving them products. The customers report them through feedback comments and others can get the right view of their products.

In the era of artificial intelligence, we can stop the displeasing assertion before expansion by correct detection of the sentiment in people's comments. The proceeding becomes more influential if we classify the sentiment more specifically. A Recurrent

Neural Network (RNN) is a type of neural network together with one of the robust DL algorithms that operate with high exactness for sentiment analysis. RNN uses the previous steps and the time in this chapter [2].

Multi-label classification is a mechanism where multiple labels can be allocated to each instance. For the classification of sentiment or the sentiment analysis more precisely, Bidirectional RNN is adopted widely to multi-labelled instances. Bidirectional Recurrent Neural Networks (Bi-RNNs) are nothing but a combination of two individualistic RNNs. They connect as hidden layers from two opposite sites.

In this chapter, we've tried to analyze sentiment with multiple classes of Bengali public online comments. This model can detect the levels of negativity as well as positivity of Bengali public comments. The aforementioned will be more effective for public acknowledgement and retreats.

The other sections of this chapter have the following: in Sect. 1.2 related works and Sect. 1.3 is narrated with the working modules. In Sect. 1.4 the experimental results have been analyzed. Finally, these exploration lines conclude with Sect. 1.5.

1.2 RELATED WORK

Md. Al- Amin et al. got the best result of 83.20% using Naive Bayes (NB) with a Bi-gram, normalizer and stammer. This model has been unable to detect multiplex Bengali statements [1]. The accuracy of 85.67% has been achieved by using RNN Bi-LSTM but the insufficient Bengali comments dataset wasn't cleaned and processed well [2]. Word2vec needs a huge dataset, but Adnan Ahmad et al. didn't narrate the dataset enough as well as got f1-score 0.91, precision 0.91 as well as recall of 0.90 [3]. The analysis of three classes of Indonesian public corpora questions dataset classification with Bidirectional-LSTM was accuracy 91% [4]. Victoria Ikoro et al. analyzed tweets of US Company Consumers and the Hu and Liu opinion lexicon has been combined for obtaining detection of frequently used topics [5]. In this typescript aspect level sentiment analysis has been implemented by SVM and NB. NB dominated in the experiment [6]. Multi Classified web page URL information has been classified with NB and got 94% exactness [7]. Al Amin et al. has analyzed Bengali sentiments by modified VADER; they preprocessed well and the model performed well too [8]. To avoid long training time, CNN got the best result with Self-Attention Gated CNN for the aspect-based sentiment [9]. Analyzing Chinese sentiments products for reviews got the best results with 93.1% accuracy with the BiGRU Attention model [10]. Word2Vec has been used to classify sentiments and got 75.5% accuracy and trained the model in six steps by the increment of 2500 comments in each step [11]. Md. Akhter-Uz-Zaman Ashik et al. has introduced a dataset of online news portal Prothom Alo's public comments on different topics [12]. They have organized the dataset and have applied CNN (accuracy 60%), SVM (accuracy 61%) and RNNs with LSTM (accuracy 75%) classification models for the evaluation. With an accuracy of 75%, RNNs with the LSTM model got the best rank of accuracy [16].

These manuscripts were reviewed and assumed that the Bengali dataset was insufficient for research. Most of the contributions based on sentiments of different

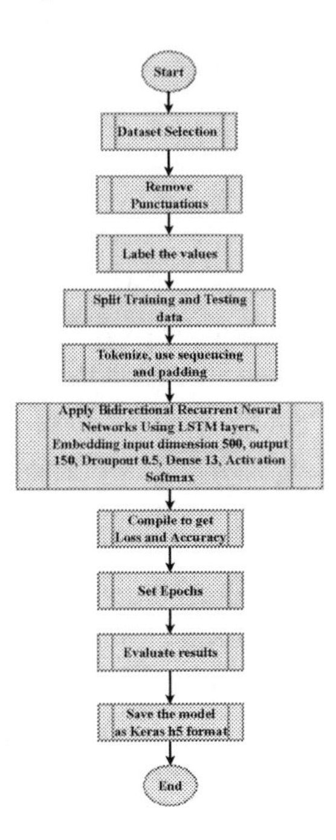

Figure 1.1 These steps of the mechanism are those we've followed in our work.

languages were reviewed. Recurrent Neural Networks (RNNs) provide satisfactory results in several multi-classified text classifications. We have tried to use Bidirectional RNNs to get better exactness than former works and tried to get multi-classified Bengali sentiment detections.

1.3 METHODOLOGY

There has been a total narration of our work procedure. The visual representation is displayed in Figure 1.1. Each step has been described in the manuscript. The methodology can be divided mainly into two fragments:

- Data preprocessing
- Bidirectional RNN Implementation

1.3.1 Data Preprocessing

Bengali benchmark dataset resource is not enough [2]. We've selected a dataset where there are 13809 entities and 2,48,562 words from the online news portal Prothom Alo's comments concerning Public Opinion, Entertainment, Lifestyle, Sports, International, Bangladesh, Education, Economy, Technology and Various fields, which were crawled and preprocessed [12]. The data set combined with polarity values $-6, -5, -4, -3, -2, -1, 0, 1, 2, 3, 4, 5$ and 6 in Figure 1.2.

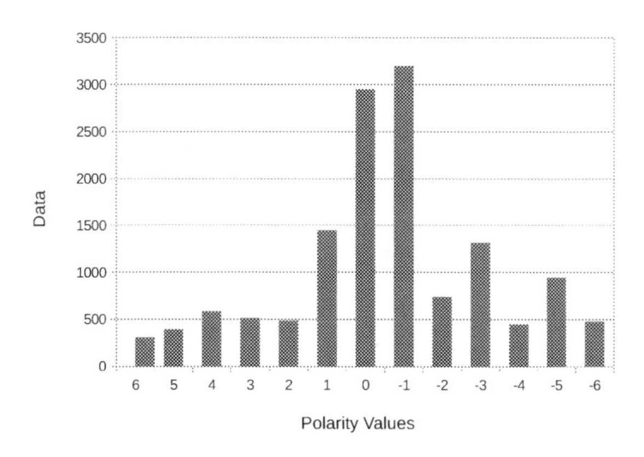

Figure 1.2 Dataset representation based on the polarity values.

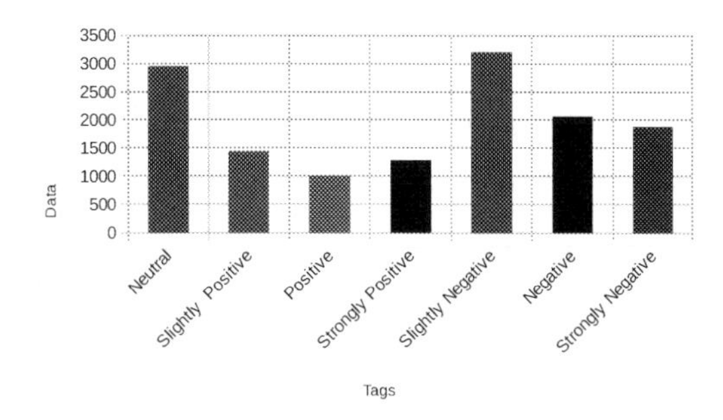

Figure 1.3 The dataset visualization after tagging the values.

For our analysis, we tagged them as Strongly Negative, Strongly positive, Negative, Positive, Slightly Negative, Slightly Positive and Neutral based on the different value ranges in Figure 1.3.

To preprocess the dataset, we removed the punctuation. Using the Label Encoder function from the Scikit-learn (Sk-learn) library, we've encoded the polarity values as categorical integer values from 0 to 12. After that, there were 13 different values shown in Table 1.1. Then the dataset has been split as 11042 for the training data set and 2767 for the testing dataset [13].

Word Embeddings (WEs) represent the words as dense word vectors. The WEs do not understand the text by the statistical structure of the language used in the dataset. The WEs target is to map semantic meaning into a geometric or embedding space. We have tokenized the data into a format for WEs. For WEs there are methods for text preprocessing and sequence preprocessing in Keras [15]. The tokenizer function tokenizer.fit_on_texts from TensorFlow (TF) has been used to tokenize the dataset with a sequential numeric value. The out of vocabulary means those are not included with the available lexicons. This OOV was sequenced. To resize them with a maximum of 500 lengths was assigned for padding. After the sequencing with the texts_to_sequences method and resizing, we've got the text sequence. Then we padded

Table 1.1 Labels of the values with tags.

Tags	Values	Labels
Strongly Negative	−6	0
	−5	1
	−4	2
Negative	−3	3
	−2	4
Slightly Negative	−1	5
Neutral	0	6
Slightly Positive	1	7
Positive	2	8
	3	9
Strongly positive	4	10
	5	11
	6	12

them with the pad_sequences function to redirect the neural network as the same size [15].

As an example, in Bengali comment data, the punctuation were removed and then the data was labelled. The texts have been converted into numeric values by the tokenizer. Then the data was sequenced using the texts_to_sequences method and text flow like [4676, 95, 275, 395, 4676, 2383, 2756] were got. After the sequencing and padding having the maximum lengths of 500 like an array length, they have been ready to pass into the Bi-RNNs.

Sk-learn and TF both provide DL tools for Data Analysis (DA) and distinctive programming for researchers and to deploy the models.

1.3.2 Bidirectional RNN Implementation

RNN, is a category of neural networks (NN), that are powerful for modelling sequence data. Feedforward NNs flow from the input to output direction but RNNs have backwards connecting points. RNNs face vanishing gradients that affect the learning of long data series. RNN with LSTM (Long Short-Term Memory) can adopt sequential contexts and remember past particulars [14]. The gradient vanishing problem is avoided by LSTM [13]. It's assumed that from the previous tasks an RNN model can perform better for text sequences [2]. To determine the next word in a sentence sequence, it is helpful to have the context, not only for just the words that come before it. In the RNN, Bidirectional copies the RNN layer passed in, and flips the back-pedal field of the newly copied layer so that it will process the inputs in alternating order [15]. Bidirectional Recurrent Neural Networks (Bi-RNNs), a special type of RNN, connect two hidden layers of reverse directions to the same output as in Figure 1.4. From the current state, the RNN can't access future input information. But the Bi-RNN can access future input information from the current state. Even in

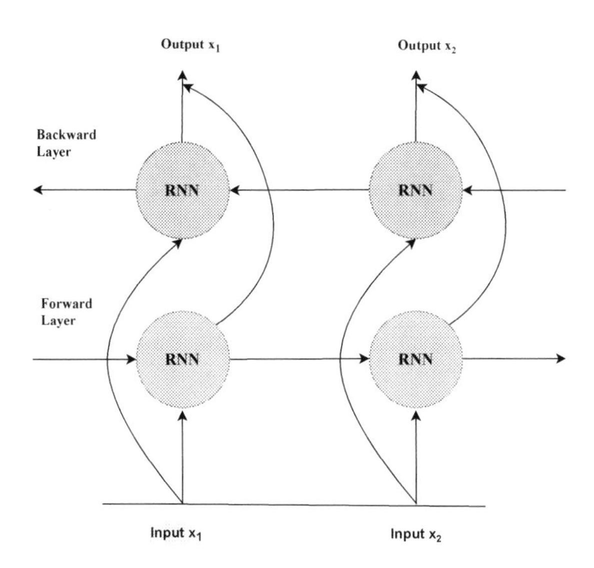

Figure 1.4 Bidirectional Recurrent Neural Networks (Bi-RNNs).

a Bi-RNN at the same state, the output layer can get information from both past and future states.

For the embedding dimension a value of 10 was determined for our dataset size initially and increased for better results. Finally, 150 was taken as an embedding dimension to acknowledge the relationship of words for better accuracy. For the embedding layer parameters, we have taken the total index size of the word and embedding dimension. Preventing overfitting, a dropout 0.5 has been applied. For the multiclass classification, loss functions are needed to find the distinction between predicted values and actual values. As we've used the actual class labels, the Sparse Categorical Cross-Entropy has been implemented. LSTM cells superintend two states of vectors. In the Bidirectional function, LSTM uses CuDNN to accelerate the performance. The dense value is 13 and softmax activation has been defined for this multi-classification [15]. After the model implementation and simulation, we saved the model as Keras h5 for the deployment.

1.4 OUTCOMES

After the 7 epochs, the results in Table 1.2 were determined through loss, accuracy, val_loss and val_acc.

Both val_loss and loss, are the cost function of cross-validation, and training data like val_acc and accuracy are the validation data and training data accuracy.

From Table 1.2, it has been assumed that the best accuracy was achieved at epoch 7 which is 88%. The result was not so bad, having better accuracy than former performances. Plotting the values of loss, accuracy, val_loss and val_acc as Figure 1.5, it has been seen that the accuracy was good but overfitting occurred. Overfitting

Table 1.2 Results of multi-labelled sentiment analysis with Bidirectional RNN.

Epochs	Loss	Accuracy	Val_loss	Val_acc
1	2.2996	0.2308	2.1561	0.2565
2	2.0928	0.2969	2.2348	0.2337
3	1.6302	0.4668	2.5026	0.1884
4	1.1514	0.6429	2.8701	0.1971
5	0.8181	0.7579	3.2375	0.1703
6	0.6038	0.8300	3.5196	0.1826
7	0.4424	0.8838	3.9765	0.1768

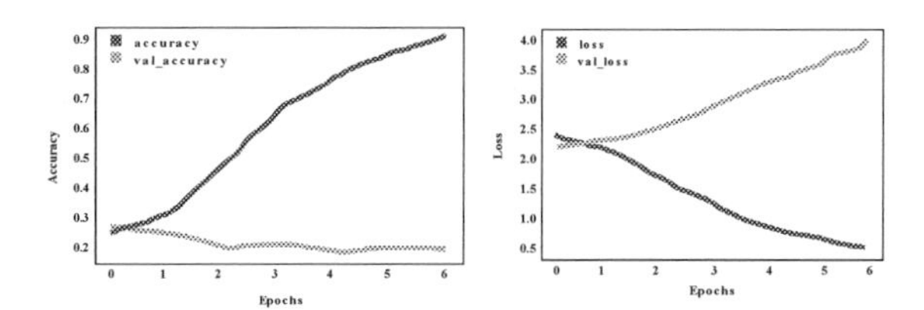

Figure 1.5 Plotted curves of loss and accuracy.

occurs if a model fits the training data too well and the loss keeps decreasing while the val_loss is increased or stable.

Finally, we've tested the model with two Bengali statements or sentences as statement -1 and statement -2 and the obtained results are in Table 1.3.

From Table 1.3, it is clear that two sentences were classified properly, because they got the highest probability distribution according to their class. Statement -1 has the

Table 1.3 Results after the testing with two Bengali sentences.

Statement - 1	Statement - 2	Labels	Tags
1.9354089e-03	7.3867358e-02	0	Strongly Negative
4.1324887e-03	1.9523818e-02	1	
1.2743475e-03	4.0217927e-03	2	
5.1211537e-04	6.0497981e-01	3	Negative
6.7992337e-02	3.3278903e-03	4	
1.0709260e-04	2.7823493e-01	5	Slightly Negative
9.7239338e-04	6.3656191e-03	6	Neutral
6.5872408e-02	8.6970031e-03	7	Slightly Positive
5.0068992e-01	2.8082356e-04	8	Positive
4.0911846e-02	2.9245173e-05	9	
2.8977847e-01	4.4753117e-05	10	Strongly Positive
2.3598161e-02	3.0431626e-04	11	
2.2230346e-03	3.2272880e-04	12	

Table 1.4 Comparison of consequences with the efficiency of the former operations.

Methodology	**Accuracy**
RNN(LSTM) Model	75%
Our proposed approach model	88%

best distribution for label 8, which is tagged positive as well as Statement -2 has the best at 3 and is tagged as negative. The exactness of our model is demonstrated in Table 1.4, which is more higher than the former accuracy [16].

1.5 CONCLUSION

Our assessment is at its very initial state. From the analysis of the results, it is clear that the training results and validation results of the assessment are overfitted to the model, which is the major weakness, but the accuracy of 88% is not bad at all. Maybe the dataset needs more prepocessing mechanisms and this dataset may not be sufficient. Currently, we are working to perceive the basis of the overfitting and trying to enhance the exactness. We are also working to deploy this approach at the production level.

Bibliography

[1] Al-Amin, M., Saiful Islam, M., & Das Uzzal S. (2017). A Comprehensive Study on Sentiment of Bengali Text. In *2017 International Conference on Electrical, Computer and Communication Engineering (ECCE)*. Cox's Bazar, Bangladesh: IEEE.

[2] Sharfuddin, A., Tihami, M., & Islam, M. (2018). A Deep Recurrent Neural Network with BiLSTM model for Sentiment Classification. In *2018 International Conference on Bangla Speech and Language Processing (ICBSLP)*. Sylhet, Bangladesh: IEEE.

[3] Ahmad, A., & Amin, M. (2016). Bengali Word Embeddings and Its Application in Solving Document Classification Problem. In *2016 19th International Conference on Computer and Information Technology (ICCIT)*. Dhaka, Bangladesh: IEEE.

[4] Anhar, R., Bharata Adji, T., & Setiawan, N. (2019). Question Classification on Question-Answer System Using Bidirectional-LSTM. In *2019 5th International Conference on Science and Technology (ICST)*. Yogyakarta, Indonesia: IEEE.

[5] Ikoro, V., Sharmina, M., Malik, K., & Batista-Navarro, R. (2018). Analyzing Sentiments Expressed on Twitter by UK Energy Company Consumers. In *2018 Fifth International Conference on Social Networks Analysis, Management and Security (SNAMS)*. Valencia, Spain: IEEE.

[6] Kag, A., Livingston, J., Livingston, M.L.M., & Livingston, L.G.X.A. (2019). Multiclass Single Label Model for Web Page Classification. In *2019 International Conference on Recent Advances in Energy-efficient Computing and Communication (ICRAECC)*. Nagercoil, India: IEEE.

[7] Vanaja, S., & Belwal, M. (2018). Aspect-Level Sentiment Analysis on E-Commerce Data. In *2018 International Conference on Inventive Research in Computing Applications (ICIRCA)*. Coimbatore, India: IEEE.

[8] Amin, A., Hossain, I., Akther, A., & Masudul Alam, K. (2019). Bengali VADER: A Sentiment Analysis Approach Using Modified VADER. In *2019 International Conference on Electrical, Computer and Communication Engineering (ECCE)*. Cox's Bazar, Bangladesh: IEEE.

[9] Yang, J., & Yang, J. (2020). Aspect Based Sentiment Analysis with Self-Attention and Gated Convolutional Networks. In *2020 IEEE 11th International Conference on Software Engineering and Service Science (ICSESS)*. Beijing, China: IEEE.

[10] Yang, L., Li, Y., Wang, J., & Sherratt, R. (2020). Sentiment Analysis for E-Commerce Product Reviews in Chinese Based on Sentiment Lexicon and Deep Learning. *IEEE Access*, 8, 23522–23530. doi: 10.1109/access.2020.2969854

[11] Al-Amin, M., Islam, M., & Das Uzzal, S. (2017). Sentiment Analysis of Bengali Comments with Word2vec and Sentiment Information of Words. In *2017 International Conference on Electrical, Computer and Communication Engineering (ECCE)*. Cox's Bazar, Bangladesh; IEEE.

[12] Ashik, M., Shovon, S., & Haque, S. (2019). *Data Set for Sentiment Analysis on Bengali News Comments*, Mendeley Data, V3, doi: 10.17632/n53xt69gnf.3

[13] Geron, A. (2019). *Hands-On Machine Learning with Scikit-Learn, Keras, and TensorFlow: Concepts, Tools, and Techniques to Build Intelligent Systems* (2nd ed., pp. 525–565). Oreilly.

[14] Ghosh, P., Azam, S., Hasib, K., Karim, A., Jonkman, M., & Anwar, A. (2021). A Performance Based Study on Deep Learning Algorithms in the Effective Prediction of Breast Cancer. In *2021 International Joint Conference on Neural Networks (IJCNN)*. IEEE.

[15] Abadi, M., Agarwal, A., Barham, P., Brevdo, E., Chen, Z., & Citro, C. et al. (2015). *TensorFlow : Large-Scale Machine Learning on Heterogeneous Systems.* Retrieved 5 July 2021, from https://www.tensorflow.org/.

[16] Ashik, M., Shovon, S., & Haque, S. (2019). Data Set for Sentiment Analysis on Bengali News Comments and Its Baseline Evaluation. In *2019 International Conference on Bangla Speech and Language Processing (ICBSLP)*. Sylhet, Bangladesh; IEEE.

CHAPTER 2

Machine Learning and Blockchain-based Privacy-aware: Cognitive Radio Internet of Things

Md Shamim Hossain

Department of Information and Communication Technology (ICT), Islamic University, Kushtia, Bangladesh

Kazi Mowdud Ahmed

Department of Information and Communication Technology (ICT), Islamic University, Kushtia, Bangladesh

Md Khairul Islam

Department of Biomedical and Engineering (BMET), Islamic University, Kushtia, Bangladesh

Md Mahbubur Rahman

Department of Information and Communication Technology (ICT), Islamic University, Kushtia, Bangladesh

Md Sipon Miah

Department of Information and Communication Technology (ICT). Islamic University, Kushtia, Bangladesh
&
Department of Signal Theory and Communications, University Carlos III of Madrid (UC3M), Leganes, Madrid, Spain
&
Internet of Vehicle[2] to Road Research Laboratory, Chang'an University, Shaanxi, China

Mingbo Niu

Internet of Vehicle to Road Research Laboratory, Chang'an University, Shaanxi, China

CONTENTS

Spectrum sensing and Dynamic Spectrum Access (DSA) in Cognitive Radio Internet of Things (CR-IoT) networks based on Cooperative Spectrum Sensing (CSS) rely on a centralized Fusion Center (FC), which is vulnerable because of the potential attacks of Malicious Users (MUs) and a single point of failure. To address this issue, this research chapter introduces a promising Machine Learning (ML) model that uses a K-means clustering algorithm to detect and cluster malicious CR-IoT users and a blockchain verification protocol as a function for secure spectrum sharing in CR-IoT networks. Each CR-IoT user in the proposed CR-IoT network appears as a sensing node and a mining node in the blockchain protocol, which leads to energy consumptions. Simulations checked the effectiveness of the suggested K-means clustering algorithm and blockchain-based privacy-aware protocol in CR-IoT networks.

2.1 INTRODUCTION

The usage of the radio spectrum has been seen as a major source of massive struggle in recent times. Demand for the radio spectrum will continue to fuel the explosive growth of new technologies like self-driving vehicles, distant healthcare, and virtual and visual reality. The Federal Communications Commission (FCC) declared that the usage of the licensed radio spectrum is comparatively limited, varying from just 15% to 85%. The remaining 15% of the licensed radio spectrum remains unused. So the fixed radio spectrum is a significant impediment for advanced wireless networking technology. In order to overcome this problem, the Internet of Things-based Cognitive Radio network is an optimistic technology [1].

The Cognitive Radio Networks (CRNs i.e., CR-IoT networks) have provided underutilized licensed spectrum to Secondary Users (SUs) i.e., CR-IoT users opportunistically [8, 17, 3]. In CR-IoT networks, the CR-IoT users perform spectrum sensing without the primary network cooperation to explore the licensed spectrum opportunities. Reliable spectrum sensing based on a single CR-IoT user is not workable because of multipath fading as well as shadowing and noise uncertainty. The CSS in CR-IoT systems is a rational solution that enlarges the usage of idle radio spectrum bands by permitting CR-IoT users [17]. Compared to multipath fading, shadowing, and receiver uncertainty troubles, Akyildiz et al. [2] provided an investigation in which spectrum sensing performance is enhanced using the CSS approach. However, the CSS is still unreliable because of the appearance of some malicious CR-IoT users. They address false radio spectrum sensing results to the Fusion Center (FC) regardless of the local radio spectrum sensing result [11]. In such cases, a few malicious CR-IoT users always report the active state of the Primary User (PU) (called Always Yes Malicious CR-IoT User (AYMU)), a few always report the inactive state of PU (called Always No Malicious CR-IoT User (ANMU)), and a few report the opposite state of PU (called Random Malicious CR-IoT User (RMU)). The existing CSS schemes are developed based on the existence of the FC, which combines the sensing outcomes from each CR-IoT user and saves them as a central database. However, the FC and the centralized database are susceptible to a single factor of failure, such as malfunctioning, because of either its own technological failures or malicious attacks. Specifically, interference may arise if the malicious user sends false sensing reports and the preceding sensing record may be tampered with in support of the attackers while the FC is hacked. In such a case, CR-IoT user(s) access to the spectrum will be harmed in this scenario along with the primary networks service and stability. To mitigate these threats, a malicious user classifier and a distributed CSS result storage are needed, rather than a centralized FC. However, in order to secure spectrum sensing and access, various researchers have used Machine Learning (ML) and a blockchain-enabled Dynamic Spectrum Access (DSA) framework. Spectrum transactions are registered and preserved in an irreversible and verifiable way for all CR-IoT users through blockchain. Shah et al. [21] developed a trustworthy ML-based CSS in CRNs, in which the FC used a weight-based combination policy to make a comprehensive decision. However, the detection of MUs and secured spectrum access were not analyzed. To enhance the system throughput by transmitting packets of different buffers through multiple channels, Zhu et al. [23] introduced a modern Q-learning-based transmission scheduling system using deep learning for the CR-IoT networks. However, the detection of MUs and secured spectrum access were not analyzed. Azmat et al. [4] recommended the use of both supervised and unsupervised ML methods for spectrum occupancy modeling that was used in all CR activities, such as spectrum management, spectrum agreement, and spectrum sensing. However, the detection of MUs and secured spectrum access were not analyzed. Jan et al. [10] proposed an SVM-based CR-IoT network for maximizing spectrum sensing with multi-class assumptions, in which the assumptions are obtained based on the quantized regions of received signal strength and the discrete levels of residual strength in a node battery. However, the detection of MUs and secured spectrum access were not analyzed. Li et al. [13] proposed an SVM-based user grouping method for CSS to reduce cooperation overhead and effectively group abnormal, redundant, and optimized users. However, the secured spectrum access was not analyzed for CR-IoT networks. Miah et al. [15] improved the throughput in the Cluster-Based Cognitive Radio Relay Network (CCRRN) using the sequential procedure for the future IoT considering a noisy reporting channel. However, the detection of MUs and secured spectrum access were not analyzed for CCRRN. Using the decaying pattern of a sparse coding procedure-based ML algorithms, Furqan et al. [7] detected the PU emulation attacks and jamming attacks in CRNs. However, the detection of MUs and secured spectrum access were not analyzed. Miah et al. [16] recommended a multi-user multiple-input and multiple-output based CR-IoT network with weighted-eigenvalue detection to reduce the spectrum scarcity, boost system throughput, extend the expected lifespan, alleviate power expenditure, and reduce the feasibility of failure. However, the detection of MUs and secured spectrum access were not analyzed. Shamim et al. [9] proposed an ML-based SVM algorithm for reliable spectrum sensing in CR-IoT networks by analyzing and classifying MUs. However, they have not analyzed secured spectrum access for CR-IoT networks. In [12], the authors suggested a puzzle-based auction-based blockchain-enabled DSA framework, where CR-IoT users are used as both sensing nodes in the CR-IoT network and mining nodes in the blockchain network. Without the use of an auctioneer, a PU shares its idle frequency band with the CR-IoT users via a blockchain. However, they have not considered the detection of MUs. To escape the collision assaults of malicious users in collaborative spectrum sensing, Xie et al. [22] proposed a blockchain-based dynamic and static reputation mechanism for secure wide band spectrum sensing. However, they have not considered secured spectrum access in CR-IoT networks. Rathee et al. [19] implies a decision-making process and blockchain-enabled CR-IoT networks which prevents data alteration by malicious users during spectrum sensing and provides security during information transmission by sensing the channel. However, they have not considered the secured spectrum access. The blockchain-based consensus multi-access mechanism is shown in [20], wherein access requests for CR-IoT users are saved as transactions and followed by consensus that is distributed among all CR-IoT users by a pre-defined law. If all CR-IoT users agree on the scheme, it still even depends on an external source for spectrum occupancy details. When a CR-IoT user does not want to access the spectrum, there is a scarcity of manifests for it to engage in CSS, which leads to energy-consuming. To address this energy-consuming problem, in [5], the authors suggested a blockchain-enabled DSA framework that enhances overall sensing energy efficiency by installing multiple sensing nodes. However, they have not considered the detection of MUs. Pei et al. [14] suggested a blockchain-based spectrum sensing policy to assure the security and decency of the spectrum transaction process. In this study, they used many SUs with sensors to provide reliable spectrum sensing. However, they have not considered the detection of MUs in CR-IoT networks. [6] described and analyzed a technique that uses a distributed anomaly detection procedure to determine falsified sensors (i.e., SUs) and record all the individual acts of SUs on the blockchain.

In this chapter, we propose an ML and blockchain enabled reliable CSS and secured DSA framework. K-means clustering is a popular ML algorithm that can group MUs and provide reliable spectrum sensing. Blockchain is a distributed ledger technology that allows CR-IoT users to update and exchange data without relying on central control and coordinator. In order to get reliable spectrum sensing and secured spectrum access, we suggest an ML and blockchain-enabled DSA framework that includes a six-phase-time-slotted protocol function. Each CR-IoT user in the proposed scheme serves as a sensing node, a miner, and verifier node. Instead of using FC, the proposed blockchain technology stores spectrum sensing, mining, and accessing data in a distributed and stable process. In [18], it is shown that spectrum sensing and mining require extra energy, and storing spectrum sensing and access data requires extra storage room. To address these issues, the ML part of our suggested framework reduces energy consumption and storage space requirements by eliminating MUs and motivates only normal users to take part in such operations. We rewarded normal CR-IoT users with tokens in the blockchain part. The tokens collected through the CR-IoT users will bid for getting the right access to the idle spectrum. So the major contributions of this chapter are summarized:

* We propose a distributive operation of reliable CSS and secured spectrum access, eliminating MUs, and heading off single factor failure because of the clustered architecture of FC.
* To improve the reliable spectrum sensing, we propose a K-means clustering ML algorithm in the CR-IoT network for grouping MUs.
* In order to improve energy efficiency motivates only normal CR-IoT users to take part in both sensing and mining acts.
* The sensing data is kept on the blockchain, which will outline the primary networks spectrum utilization and consequently serve them an extra approach for network management and protection. Which could allow many more licensed users to engage in spectrum sharing.

The remainder of the chapter is organized as follows: The system model is addressed in Section 2.2 in terms of cognitive networks and blockchain networks. Implementation of the sensing-clustering-biding-mining strategy discussed in Section 2.3. The findings of the simulation are presented in Section 2.4. Finally, we draw our conclusion and describe future works in Section 2.5. Moreover, Table 2.1 outlines the parameters that are used in this simulation.

2.2 SYSTEM MODEL

The proposed system model is described in the following subsections 2.2.1 and 2.2.2.

2.2.1 *Blockchain based CR-IoT Network*

To determine the PUs activities, we considered N normal and M malicious CR-IoT users in the proposed CR-IoT network, where each CR-IoT user performs spectrum sensing and reports the sensing results to the blockchain network. The received sensing information from MUs is always higher or lower than normal users (i.e., applies to the active state or inactive state of PUs), which degrades the performance of CR-IoT networks. After receiving sensing information, the proposed K-means clustering ML

Table 2.1 The summary of parameters used in this chapter.

Parameters	Descriptions
N	Normal CR-IoT users
M	Malicious CR-IoT users
B	Prior bandwidth in kHz
τ	Sensing time duration
T	Total received samples
Pr_s	Sensing probability
Pr_m	Mining probability
Pr_f	False alarm probability
Pr_0	Channel idle probability
λ	Packet arrival rate
τ_s	Sensing token reward
τ_m	Mining token reward
ε_s	Sensing energy consumption
ε_m	Mining energy consumption

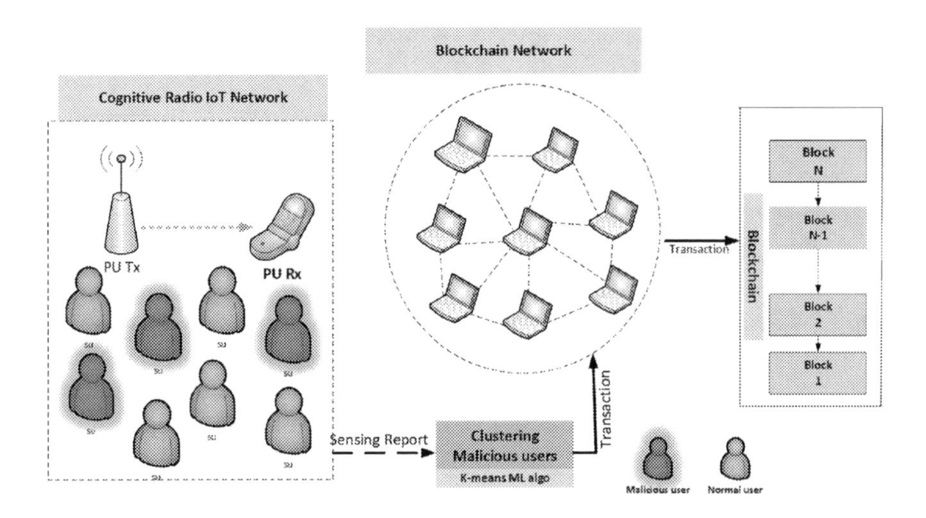

Figure 2.1 Blockchain based CR-IoT network system.

algorithm will group the MUs. For spectrum management, blockchain acts as a stable archive, where spectrum sensing data, spectrum auction results, spectrum access history, and unoccupied frequency bands all will be stored. The CR-IoT user(s) will access the temporary idle channel belonging to a primary network via blockchain based on normal CR-IoT user information. The overall proposed system model is shown in Figure 2.1.

To estimate the approximate PU activities, define the ith signal sample taken by the uth CR-IoT user as $z_u(i)$:

$$z_u(i) = \begin{cases} n_u(i); \\ h_u(i)\,s(i) + n_u(i); \end{cases} \tag{2.1}$$

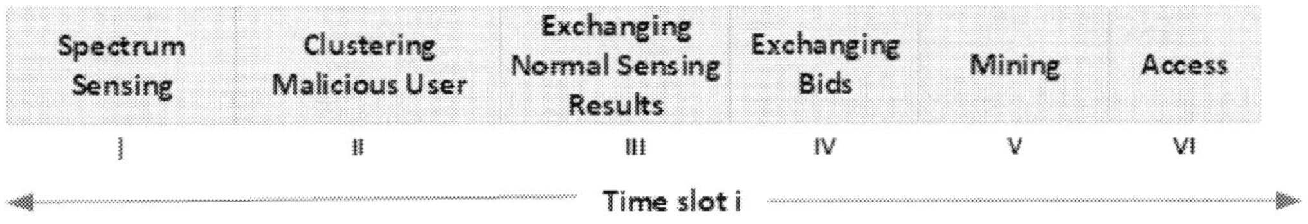

Spectrum Sensing	Clustering Malicious User	Exchanging Normal Sensing Results	Exchanging Bids	Mining	Access
I	II	III	IV	V	VI

Time slot i

Figure 2.2 The operations sequence protocol in a time slot.

where $i = 1, 2, \ldots, T$. Here, T is the total received samples taken by the uth CR-IoT user, which is specified as $T = \tau B$, τ is the sensing time period in seconds, and B is a prior known bandwidth in kHz. Moreover, $s(i)$ is the transmitted signal by the PU with zero mean μ_0 and variance σ_0^2, that is, $s(i) \sim \aleph\left(\mu_0, \sigma_0^2\right)$ and $n_u(i)$ is a circularly symmetric complex Gaussian (CSCG) noise with variance $\sigma_{n,u}^2$ of the uth CR-IoT user, i.e., $z_u(i) \sim \aleph(0, \sigma_{z,u}^2)$. Also, $h_u(i)$ is the channel gain of the uth CR-IoT user.

The estimated energy level (E_u) of the uth CR-IoT user from the received signal samples is defined as follows [16]:

$$E_u = \frac{1}{T} \sum_{i=1}^{T} |z_u(i)|^2 \tag{2.2}$$

2.2.2 *The Protocol Structure*

In the proposed system model, we consider a six-phase-time-slot operations protocol to ensure reliable CSS and secure spectrum access as shown in Figure 2.2.

1. **Sensing (Phase I):** Every CR-IoT user must determine whether to sense or rest at the start of every frame. Then whoever wants spectrum sensing, performs spectrum sensing and reports sensing data to determine the activities of PUs. CR-IoT users who take part in the CSS will be rewarded with τ_s.

2. **Clustering MUs (Phase II):** After receiving spectrum sensing information, the proposed K-means clustering algorithm will cluster the MUs, who send false sensing reports.

3. **Exchanging Normal Sensing Results (Phase III):** Each normal CR-IoT user who has conducted spectrum sensing, exchanges their finding outcomes with all other CR-IoT users and determines the global PU activities based on the majority rule.

4. **Exchanging Bids (Phase IV):** When the global decision shows that the channel is inactive, every CR-IoT user who wishes to use the spectrum band makes a bid for access according to their bidding policies and exchanges it.

5. **Mining (Phase V):** All the CR-IoT users determine whether to operate a mining operation in accordance with its mining strategy. The active miner will build and publish a new block that contains the sensing results, CR-IoT users bids, and the successful CR-IoT user(s). The mining winner CR-IoT users will be rewarded with τ_m.

6. **Access (Phase VI):** Finally, the leading CR-IoT user receives an acknowledgment (ACK) to access the spectrum. If the auction results in a tie, the leading bidders will share the frequency band using Time Division Multiple Access (TDMA).

2.3 SENSING-CLUSTERING-BIDING-MINING POLICY

Based on the proposed DSA framework, four policies are essential for CR-IoT users to opportunistically access the idle channel. Each CR-IoT user must need to sense and mine in all time slots in order to maximize token revenues. But, it might be a time- and energy-intensive operation. We recommend a set of policies for CR-IoT users to address these concerns.

1. **Sensing Policy:** A sensing policy $a_u(i)$ determines when a CR-IoT user needs to sense the channel. Let the sensing policy $(a_u(i))$ of the uth CR-IoT user at the ith time slot with sensing probability Pr_s be defined as [18]:

$$a_u(i) = \begin{cases} 1, & \text{for } Pr_s; \\ 0, & \text{for } 1 - Pr_s; \end{cases} \tag{2.3}$$

   where $a_u(i) = 1$ and $a_u(i) = 0$ represents that the uth CR-IoT user requires sensing the channel and does not require sensing the channel, respectively.

2. **Clustering Policy:** A clustering policy determines how malicious CR-IoT users separate from the CR-IoT network. K-means clustering is a technique that allows us to find groups of similar CR-IoT users. In this article, we consider that N normal and M malicious (total $U = N + M$) CR-IoT users sense the channel. In addition, the received sensing information data D is expressed as:

$$D = \{x_1, x_2, x_3, .., x_u, .., x_U\} \tag{2.4}$$

   where x_u denotes the sensing information of the uth CR-IoT user. After receiving sensing information, the sensing dataset D is applied to the proposed K-means algorithm to separate MUs from the CR-IoT networks.

   The K-means algorithm requires a predetermined number of clusters to categorize CR-IoT users into similar groups. The elbow method is an effective way to figure out how many clusters are needed to categorize the CR-IoT users. The elbow method is nothing more than a distortion function known as the Sum of Square Error (SSE). The SSE provides the optimum number of clusters in a dataset D by measuring the distance between each CR-IoT user of the cluster and its centroid.

$$Distortion(K) = \sum_{j=1}^{K} \sum_{u \epsilon c_j} ||x_u - c_j||^2 \tag{2.5}$$

To provide the cluster number, we initialize K random cluster centroids and calculate the SSE among the centroids and CR-IoT users data. For any K clusters when the distortion remains constant even if we increase the value of K – that's the achieved K is the number of clusters. The pseudo-code of the K-means algorithm is given below:

(a) Initialize K random cluster centroids (i.e.,$\{c_1, c_2, .., c_k\}$).

(b) Assign each CR-IoT user x_u to their closest cluster centroid using the Euclidean distance function:

$$C_k = Cluster(x_u) = min \sum_{u=1}^{U} \sum_{k=1}^{K} ||x_u - c_k||^2 \tag{2.6}$$

(c) Calculate the centroid or mean of all CR-IoT users in each cluster and move the cluster centroids to that mean value. Now we have new cluster centroids.

(d) Repeat b and c until the same centroid is assigned to each cluster in consecutive rounds.

3. **Biding Policy:** A bidding policy, $b_u(i)$, determines how many tokens CR-IoT users should bid for spectrum access. Let the bidding policy ($b_u(i)$) of the uth CR-IoT user at i^{th} time slot is defined as [18]:

$$b_u(i) = \frac{q_u(i)}{Q_u} n_u(i) \tag{2.7}$$

where $q_u(i)$, and Q_u denote the number of buffers and buffer size of the uth CR-IoT user, respectively. $n_u(i)$ denotes the current maximum number of available tokens of uth CR-IoT user, i.e., $b_u(i) \leq n_u(i)$.

4. **Mining Policy:** A mining policy ($c_u(i)$) determines when CR-IoT users should work on mining to update the blockchain. Let the mining policy $c_u(i)$ of the uth CR-IoT user at the ith time slot with Pr_m mining probability is defined as [18]:

$$c_u(i) = \begin{cases} 1, & \text{for } Pr_m; \\ 0, & \text{for } 1 - Pr_m; \end{cases} \tag{2.8}$$

where $c_u(i) = 1$ and $c_u(i) = 0$ represent mining and not mining, respectively.

2.3.1 Sensing-Mining Energy Efficiency

The CSS will be the most reliable and the blockchain will be consistently updated when CR-IoT users sense and mine at all times. As a result, this will lead to energy consumption. To overcome this problem, we extract the energy efficiency performance of the proposed CR-IoT networks based on the sensing-mining policy. The sensing and mining techniques are both random operations with defined probabilities, i.e., Pr_s and Pr_m.

Let the probability of spectrum sensing of n CR-IoT users that are taking part in spectrum sensing in a time period be defined as follows [18]:

$$Pr_{\text{sensing },n} = \binom{N}{n} (Pr_s)^n (1 - Pr_s)^{(N-n)} \tag{2.9}$$

If there is at least one CR-IoT user working as a miner in a time slot, the blockchain will be revised. Thus, the probability of updating the blockchain is:

$$Pr_{\text{mining }} = 1 - (1 - Pr_m)^N \tag{2.10}$$

When at least one CR-IoT user acts as a miner to update the blockchain it confirms that there was no falsification result in the CSS. Then the CR-IoT user(s) can access the channel if it is idle. The chance of accessing the channel in the specific time period of the winning CR-IoT user(s) is:

$$Pr_{\text{access }} = \sum_{n=0}^{N} Pr_{\text{sensing },n} (1 - Pr_{f,n}) Pr_{\text{mining }} Pr_0 \tag{2.11}$$

where $Pr_{f,n}$ denotes the identical false alarm probability of n CR-IoT users who took part in the spectrum sensing, which is defined as follows:

$$Pr_{f,n} = \sum_{j=\lceil \frac{n}{2} \rceil}^{n} \binom{n}{j} Pr_f^j (1 - Pr_f)^{n-j} \tag{2.12}$$

However, CR-IoT users continuously engage in sensing and mining, which will lead to energy consumption. Here, additional energy is required to enforce the proposed framework. The energy wastage for sensing and mining is denoted by ε_s and ε_m respectively. Thus the energy efficiency of sensing-mining policy is mathematically described as:

$$\eta = \frac{Pr_{access} * P}{N (Pr_s \varepsilon_s + Pr_m \varepsilon_m)} \tag{2.13}$$

where P denotes the maximal number of packets transmitted by a CR-IoT user during a time slot.

2.4 SIMULATION RESULTS AND DISCUSSION

In this segment, we execute a comprehensive simulation using a heuristic sensing-clustering-biding-mining policy to analyze the overall performance of the proposed ML and blockchain-enabled DSA scheme.

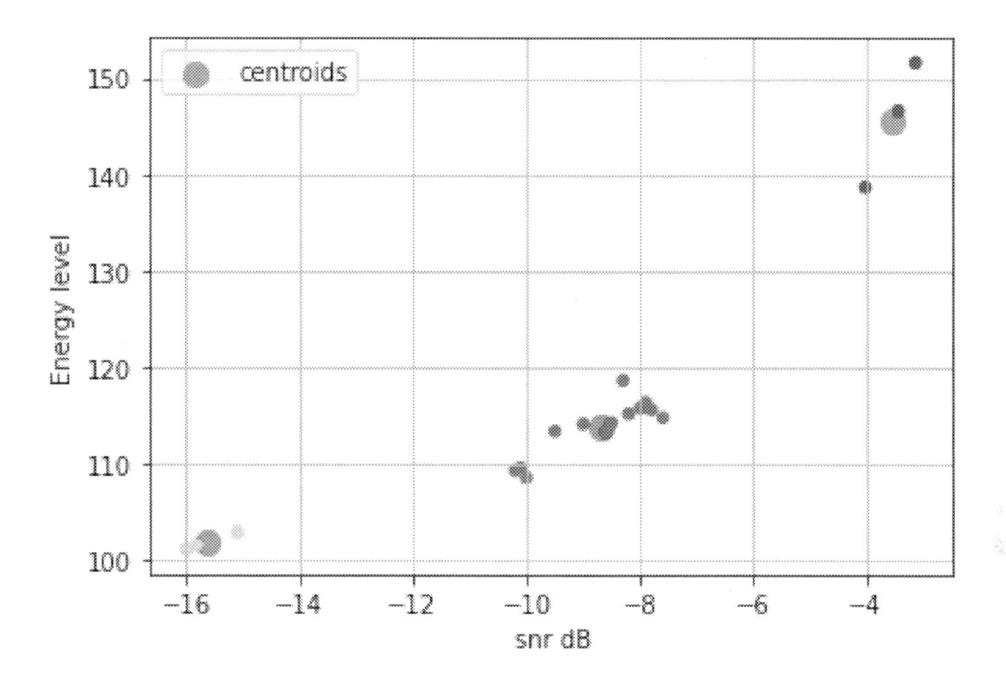

Figure 2.3 Clustering of malicious users and normal CR-IoT users.

To execute our simulation, we examine the proposed system model with $M = 6$ MUs and $N = 14$ normal users (total $U = 20$ users). To estimate the energy level vector of each CR-IoT user, we take $T = 1000$ signal samples with 1000 iterations. The SNR range of each CR-IoT user is considered between -17dB to -3dB. For flexibility, we split our simulation into two parts.

First, we used a K-means clustering ML algorithm to detect and classify MUs from the CR-IoT network. Figure 2.3 shows the clustering of normal CR-IoT users and MUs. It shows that our proposed K-means algorithm clusters the CR-IoT users into three groups. A few CR-IoT users notice high sensing reports, which are noted as AYMUs. A few CR-IoT users notice low sensing reports, which are noted as ANMUs. Some CR-IoT users notice normal sensing reports, which are noted as normal users.

Secondly, we execute the effectiveness of our submitted K-means algorithm for the energy efficiency of sensing-mining. We consider an equivalent packet arrival rate λ for all CR-IoT users. In such a case, $P = \frac{N\lambda}{Pr_0}$, and here P is the channel capacity. To inspect the actuality of the malicious users on the execution, we fluctuate the sensing chance Pr_s (varied from 0 to 1) with the constant mining chance $Pr_m = 1$. Figure 2.4, depicts the energy efficiency of sensing-mining at two separate packet arrival rates (i.e., $\lambda = 4$, and $\lambda = 3$), when malicious users exist in the CR-IoT network and after malicious users are excluded from the CR-IoT network. According to Figure 2.4, the sensing-mining energy efficiency does not improve flatly with the expansion of sensing chance, and there is a gap in sensing-mining energy efficiency when malicious users are present in the CR-IoT network. When malicious users are present in the CR-IoT network, the energy efficiency of sensing-mining is reduced.

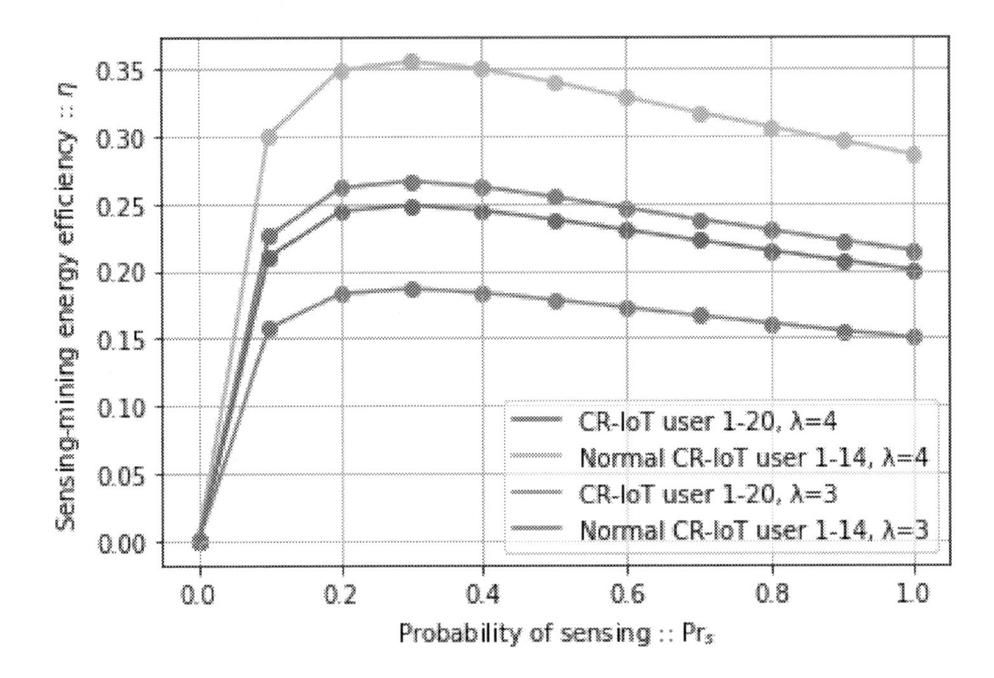

Figure 2.4 Sensing-Mining energy efficiency η vs probability of spectrum sensing Pr_s when malicious users exist and after clustering with probability of mining $Pr_m = 1$.

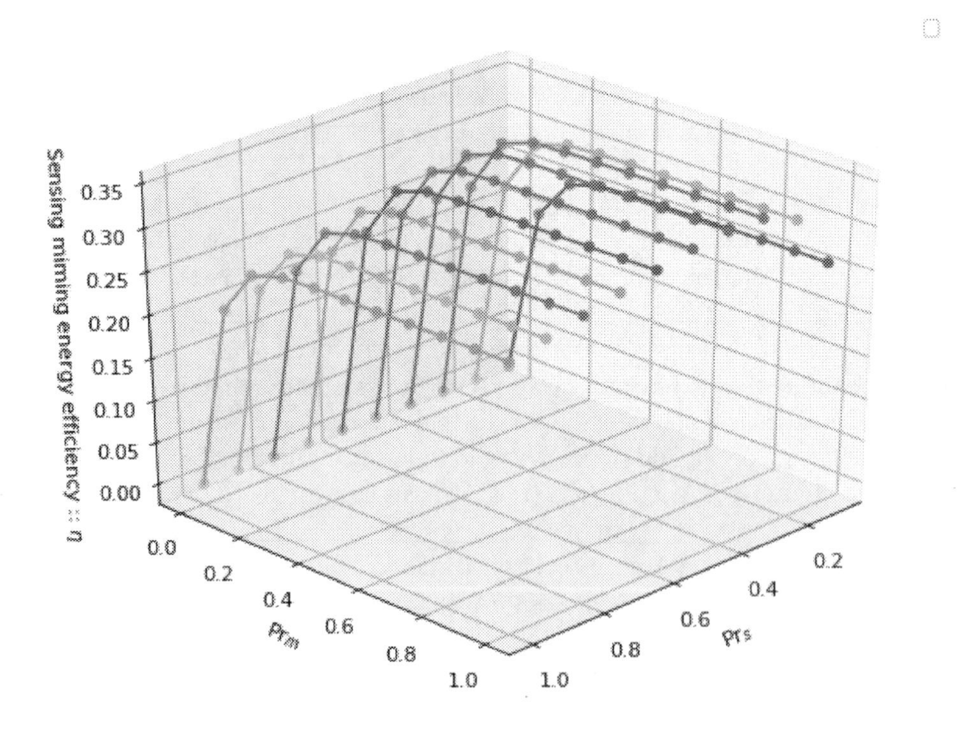

Figure 2.5 Sensing-mining energy efficiency, η vs. probability of sensing, Pr_s, and probability of mining, Pr_m.

Furthermore, we observe the effect of both sensing and mining probability (varied from 0 to 1) on the energy efficiency of sensing mining in Figure 2.5. It exhibits the energy efficiency behavior of sensing mining with $\lambda = 4$, and $P = 48$. It appears that the sensing mining energy efficiency result doesn't increase regularly with a rise of sensing and mining probability prospects.

2.5 CONCLUSION

Machine learning and blockchain-enabled DSA architecture is a promising technique for reliable CSS and secure spectrum access in CR-IoT networks. However, there remain a few limitations, such as transaction cost and latency. Transaction costs include publishing a new block, and verification by other users. Latency includes the mining process and updating the local blockchain of all users. In this chapter, we have proposed a six-phase-time-slotted protocol function, to motivate normal CR-IoT users to participate in sensing and mining operations. In this proposed DSA framework, to achieve reliable spectrum sensing and secure spectrum access, normal CR-IoT users act as both sensing nodes and mining nodes. Furthermore, normal CR-IoT users participate in such sensing mining energy consumption operations.

ACKNOWLEDGMENT(S)

This research was supported, in part, by the Islamic University (IU), Kushtia-7003, Khulna, Bangladesh (Ref. No. 141/EDU/IU-2020/634) and in part, by Project PCN No.IC20201325, funded by Bangladesh Bureau of Educational Information & Statistics (BANBEIS), Ministry of Education, Government of the People's Republic of Bangladesh.

DISCLOSURE STATEMENT

The authors declare that there is no conflict of interest that could have published of this chapter.

Bibliography

[1] Miah, M., Ahmed, K. M., Islam, M., Mahmud, M., Raihan, A., Rahman, M., & Yu, H (2020). Enhanced sensing and sum-rate analysis in a cognitive radio-based internet of things. *Sensors 20.9* (2020): 2525.

[2] Akyildiz, I. F., Lo, B. F., & Balakrishnan, R. (2011). Cooperative spectrum sensing in cognitive radio networks: A survey. *Physical Communication*, 4(1), 40–62.

[3] Amin, M. R., Rahman, M. M., Hossain, M. A., Islam, M. K., Ahmed, K. M., Singh, B. C., & Miah, M. S. (2018). Unscented Kalman filter based on spectrum sensing in a cognitive radio network using an adaptive fuzzy system. *Big Data and Cognitive Computing*, 2(4), 39.

[4] Azmat, F., Chen, Y., & Stocks, N. (2015). Analysis of spectrum occupancy using machine learning algorithms. *IEEE Transactions on Vehicular Technology*, 65(9), 6853–6860.

[5] Bayhan, S., Zubow, A., & Wolisz, A. (2018, October). Spass: Spectrum sensing as a service via smart contracts. *In 2018 IEEE International Symposium on Dynamic Spectrum Access Networks (DySPAN)* (pp. 1–10). IEEE.

[6] Careem, M. A. A., & Dutta, A. (2019, November). Sensechain: Blockchain based reputation system for distributed spectrum enforcement. *In 2019 IEEE International Symposium on Dynamic Spectrum Access Networks (DySPAN)* (pp. 1–10). IEEE.

[7] Furqan, H. M., Aygül, M. A., Nazzal, M., & Arslan, H. (2020). Primary user emulation and jamming attack detection in cognitive radio via sparse coding. *EURASIP Journal on Wireless Communications and Networking*, 2020(1), 1–19.

[8] Hossain, E., & Bhargava, V. K. (Eds.). (2007). Cognitive Wireless Communication Networks. *Springer Science & Business Media.*

[9] Hossain, M. S., & Miah, M. S. (2021). Machine learning-based malicious user detection for reliable cooperative radio spectrum sensing in Cognitive Radio-Internet of Things. *Machine Learning with Applications*, 100052.

[10] Jan, S. U., Vu, V. H., & Koo, I. (2018). Throughput maximization using an SVM for multi-class hypothesis-based spectrum sensing in cognitive radio. *Applied Sciences*, 8(3), 421.

[11] Kaligineedi, P., Khabbazian, M., & Bhargava, V. K. (2010). Malicious user detection in a cognitive radio cooperative sensing system. *IEEE Transactions on Wireless Communications*, 9(8), 2488–2497.

[12] Kotobi, K., & Bilen, S. G. (2018). Secure blockchains for dynamic spectrum access: A decentralized database in moving cognitive radio networks enhances security and user access. *IEEE Vehicular Technology Magazine*, 13(1), 32–39.

[13] Li, Z., Wu, W., Liu, X., & Qi, P. (2018). Improved cooperative spectrum sensing model based on machine learning for cognitive radio networks. *IET Communications*, 12(19), 2485–2492.

[14] Lv, P., Zhao, H., & Zhang, J. (2020, December). Blockchain Based Spectrum Sensing: A Game-Driven Behavior Strategy. *In 2020 IEEE 9th Joint International Information Technology and Artificial Intelligence Conference (ITAIC)* (Vol. 9, pp. 899–904). IEEE.

[15] Miah, M. S., Schukat, M., & Barrett, E. (2018). An enhanced sum rate in the cluster based cognitive radio relay network using the sequential approach for the future Internet of Things. *Human-centric Computing and Information Sciences*, 8(1), 1–27.

[16] Miah, M. S., Schukat, M., & Barrett, E. (2020). Sensing and throughput analysis of a MU-MIMO based cognitive radio scheme for the Internet of Things. *Computer Communications*, 154, 442–454.

[17] Mitola, J., & Maguire, G. Q. (1999). Cognitive radio: Making software radios more personal. *IEEE personal communications*, 6(4), 13–18.

[18] Pei, Y., Hu, S., Zhong, F., Niyato, D., & Liang, Y. C. (2019, December). Blockchain-enabled dynamic spectrum access: Cooperative spectrum sensing, access and mining. *In 2019 IEEE Global Communications Conference (GLOBECOM)* (pp. 1–6). IEEE.

[19] Rathee, G., Ahmad, F., Kurugollu, F., Azad, M. A., Iqbal, R., & Imran, M. (2020). CRT-BIoV: A cognitive radio technique for blockchain-enabled internet of vehicles. *IEEE Transactions on Intelligent Transportation Systems.*

[20] Seo, H., Park, J., Bennis, M., & Choi, W. (2018, October). Consensus-before-talk: Distributed dynamic spectrum access via distributed spectrum ledger technology. *In 2018 IEEE International Symposium on Dynamic Spectrum Access Networks (DySPAN)* (pp. 1-7). IEEE.

[21] Shah, H. A., & Koo, I. (2018). Reliable machine learning based spectrum sensing in cognitive radio networks. *Wireless Communications and Mobile Computing*, 2018.

[22] Xie, X., Hu, Z., Chen, M., Zhao, Y., & Bai, Y. (2021). An Active and Passive Reputation Method for Secure Wideband Spectrum Sensing Based on Blockchain. *Electronics*, 10(11), 1346.

[23] Zhu, J., Song, Y., Jiang, D., & Song, H. (2017). A new deep-Q-learning-based transmission scheduling mechanism for the cognitive Internet of Things. *IEEE Internet of Things Journal*, 5(4), 2375–2385.

CHAPTER 3

Machine Learning-based Models for Predicting Autism Spectrum Disorders

S. M. Mahedy Hasan

Rajshahi University of Engineering and Technology, Rajshahi, Bangladesh

Md. Fazle Rabbi

Bangladesh Army International University of Science and Technology, Comilla, Bangladesh

Arifa Islam Champa

Bangladesh Army International University of Science and Technology, Comilla, Bangladesh

Md. Rifat Hossain

Bangladesh Army International University of Science and Technology, Comilla, Bangladesh

Md. Asif Zaman

Rajshahi University of Engineering and Technology, Rajshahi, Bangladesh

CONTENTS

A Autism Spectrum Disorder (ASD) is a category of disorders in neurodevelopment that is not remediable but can be alleviated by quick interventions. We collected ASD datasets for toddlers and adults and performed various Feature

Transformation (FT) methods, including Minmax Scaler (MS), Standard Scaler (SS), Robust Scaler (RS) on those ASD datasets. Subsequently, different ensemble classification techniques were applied to these transformed ASD datasets and their classification outcomes were analyzed. A variety of statistical measurements, such as accuracy, ROC, F1-score, precision, recall, MCC, kappa, and log loss, were considered to justify experimental findings. The experimental results implied that the AdaBoost (AB) classifier for toddlers and both the Random Forest (RF) and CatBoost (CB) for the adult's dataset showed the best performance compared with other classification methods. In contrast, the feature transformations resulting in the classification outcomes of RS for toddlers and SS for adults showed better performance. After those analyses, various Feature Selection Techniques (FST) were then considered to evaluate the risk factors of RS transformed toddlers and SS transformed adults. The findings of these meticulous procedures specify that proper optimization of machine learning techniques can play a significant role in predicting ASD.

3.1 INTRODUCTION

Autism Spectrum Disorder (ASD) is a neurodevelopmental incapacity related to brain improvement that begins in early childhood and affects a person's social interactions and communication problems. ASD has limited and repetitive patterns of behavior and the term spectrum includes an extensive range of symptoms and severity. As there is no permanent cure for ASD, only early detection and proper treatment can make a huge difference in children's lives and improve children's behavior and communication skills. However, ASD detection and diagnosis are very challenging and complicated when using conventional behavioral studies. Generally, at around two years of age, ASD is most frequently identified, but it can be recognized later, depending on the symptoms. To diagnose ASD as early as possible, there are several clinical instruments available, though in practice, these diagnostic methods are not frequently utilized, except when there is a massive peril of ASD development. Alison et al. [3] presented a brief measurable catalog that can be utilized at various periods in a patient's life, such as toddlers, children, adolescents, and adults. Later, Thabtah et al. [4] produced an ASDTests mobile application tool for detecting ASD as early as possible based on some questionnaires, Q-CHAT, and AQ-10 tools. Subsequently, they also built an open-source dataset using the mobile app's data and uploaded those datasets in the publicly available repository named the University of CaliforniaIrvine (UCI) Machine Learning Repository and Kaggle for further improvement in this research field.

In recent years, significant research has been performed using different machine learning techniques to detect and diagnose ASD as early as possible. Thabtah et al. [5] assessed the ASD features using Rule-based Machine Learning (RML) methods and affirmed that RML lets classifiers improve classification performance. Omar et al. [6] merged the Random Forest (RF) and Iterative Dichotomiser 3 (ID3) algorithm and developed a prediction model for screening children, adolescents and adults. Abbas et al. [7] developed a single assessment tool by merging ADI-R and ADOS ML techniques and employed various feature encoding methods to combat the inadequacy, sparsity, and problems with unbalanced data. Another research conducted by

Thabtah et al. [4] showed the feature-to-class and feature-to-feature correlation values using the computational intelligence method and applied Support Vector Machines (SVM), Decision Tree (DT) and Logistic Regression (LR) as classifiers for diagnosis and prognosis of ASD [8]. Furthermore, Goh et al. [9] analyzed typically developed (TD) (N=19) and ASD (N=11) patients, where a correlation-based feature selection (CFS) was employed to assess the significance of the features. Crippa et al. [10] analyzed ASD and TD children and classified 15 pre-school ASDs utilizing only seven attributes. However, they showed that cluster analysis could better seize complicated features that predict ASD phenotype and heterogeneity. Tyagi et al. [11] compared the classification performance of KNN, LR, LDA, CART, NB and SVM for ASD detection of adults. Duda et al. [12] explored the ASD data and noticed that only a total of 65 features were adequate to detect ASD from attention deficit hyperactivity disorder (ADHD). In 2019, Shuvo et al. [13] developed a Random Forest (RF) based prediction model to detect ASD using behavioral attributes. Besides this, Osman et al. [14] applied Linear Discriminant Analysis (LDA) and K-Nearest Neighbors (KNN) algorithms to classify ASD children of "4–11" years. In 2018, Fatiha et al. [15] proposed an ASD prediction model based on the RF method for children of 4–11 years. Faiz et al. [16] analyzed the classification performance of a Deep Neural Network (DNN) in diagnosing ASD using two separate datasets of adults. Hardi et al. [17] compared the performance of SVM's different kernels in classifying ASD for children's data and found that the polynomial kernel performed better. After collecting the ASD datasets we performed some initial preprocessing (handling missing values, feature encoding) of those datasets and then several FT techniques were employed to transform the datasets into a suitable format for further analysis. After that, different ensemble machine learning classification algorithms were applied to those converted datasets and the best ML approaches were identified.Besides, we also investigated how feature transformation improves the classification outcome of the ensemble ML algorithms. The remaining segment of the chapter is organized as follows: Section 3.2 describes the datasets and research methodology utilized in this study. Section 3.3 analyzes the experimental outcomes of this research. Section 3.4 concludes the research chapter.

3.2 MATERIALS AND METHODS

This section discusses the materials and methods.

3.2.1 Dataset Description

This study gathered two different ASD datasets for toddlers and adults from the publicly available repository Kaggle and the UCI Machine Learning Repository [18][19]. A detailed description of the different datasets used in this study is given below in Table 3.1

Table 3.1 Details of datasets.

Toddlers	16	1054	728	326
Adults	16	704	508	196

Table 3.2 Detailed description of different FT methods.

Min Max Scaler (MS)	It transforms all the data value between 0 and 1.
Standard Scaler (SS)	It converts the feature values in the form of mean value 0 and standard deviation 1.
Robust Scaler (RS)	It uses interquartile range to transform the feature values and handle the outliers.

3.2.2 Methods

In this research, two different ASD datasets (toddlers, adults) were employed to create prediction models. At first, the datasets were checked for missing values, and K-Nearest Neighbors (KNN) imputation was used to fill the missing values. Afterwards, One-Hot Encoding was used to turn the dataset's categorical feature values into integer values. All the datasets used in this research had imbalanced class distribution problems, so the oversampling method named random over sampler was used to solve this problem. Thus, three standard FT techniques, such MS, SS, RS were then used in those datasets (see details in Table 3.2). The transformed datasets were then classified using nine (9) different ensemble classification methods. Moreover, we analyzed the classification performance and identified the best performing classifiers and FT methods for each dataset. Finally, the ASD risk factors were calculated and the most important features were ranked by using two different FSTs, such as Info Gain Attribute Evaluator (IGAE) and Correlation Attribute Evaluator (CAE) (see details in Table 3.3). Figure 3.1 shows the sequential workflow of this research methodology, how we investigated and examined the risk factors of ASD.

Table 3.3 Detailed description of different FST methods.

Info Gain Attribute Evaluator (IGAE)	It calculates the information gain value or entropy for each feature with respect to the output variable.
Correlation Attribute Evaluator (CAE)	It assesses the quality of a feature by calculating the Pearson's correlation value between it and the output variable.

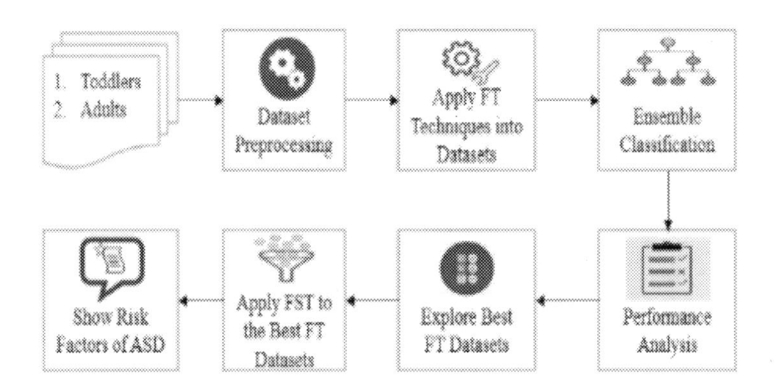

Figure 3.1 Sequential workflow to detect autism spectrum disorders.

3.2.3 Classification Techniques

AdaBoost (AB): AdaBoost is an ensemble classifier that combines multiple weak classifiers to create a robust classifier. It chooses the training set and assigns the weights iteratively for retraining the algorithm based on previous training accuracy. An arbitrary subset of the overall training set is used to teach every poor classifier, and after training, AB assigns weight to each training item as well as the classifier.

Gradient Boosting Machine (GBM): GBM is an ensemble ML algorithm that combines multiple weak learning models to produce a robust prediction model. Generally, decision trees are used as weak learners and each time a weak learner is added to the model, the previous learner's weights are frozen.

Histogram-Based Gradient Boosting (HB): HB is an ensemble ML algorithm and comes from the GBM family. HB works by using a data structure called a histogram, where the instances are entirely ordered. The input data X needs to be binned into integer-valued bins to construct a histogram. In order to complete the binning procedure, the feature values need to be sorted, but it is required only once at the beginning of the boosting process.

Extreme Gradient Boosting Machine (XGB): XGB is a popular variant of GBM that employs a gradient boosting structure. It creates the decision trees in sequential manner and weights are assigned to the independent variables and then fed in to the decision tree, which predict results.

Light Gradient Boosting Machine (LM): Light GBM stands for a lightweight gradient boosted machine based on decision trees, and is a free, open-source gradient boosting framework provided by Microsoft. However, Gradient-based One-Side Sampling (GOSS) is used to separate out data samples for obtaining a split value.

CatBoost (CB): CB is a fully accessible machine learning technique focused on gradient boosting decision trees and developed by Yandex researchers. Unlike other gradient boosting algorithms, it implements symmetric trees during the creation of decision trees. During training, a set of symmetric decision trees is built sequentially. Each successive tree is built with reduced loss (logarithmic loss) compared to the previous trees.

Table 3.4 Detailed description of evaluation measures.

Name	Formula
Accuracy	Accuracy =(TN + TP)/(TN + TP + FN + FP)
ROC	TPR=TP/(TP + FN) FPR = FP/(FP + TN)
F1-Score	F1-Score = 2TP / (FN + FP + 2TP)
Precision	Precision = TP/ (FP + TP)
Recall	Recall =TP /(FN + TP)

Random forest (RF): RF is an ensemble ML algorithm that follows the divide-and-conquers approach to build multiple decision trees from the input dataset. It generally works in two steps: first, it combines multiple decision trees to form a forest, and second, it makes predictions from the trees created in the first step.

Extra Trees (ET): ET is an extremely randomized tree that is an ensemble ML algorithm that generates multiple unpruned decision trees using the training set. Unlike the RF algorithms, the whole learning sample is used to train each decision tree, and secondly, the tree learner's top-down break is randomized. The majority voting technique is used to make predictions where each decision tree gives a vote, and the highest voted prediction is considered the final classification result.

Bagging (BG): Bootstrap aggregation is also known as bagging, a subclass of ensemble ML algorithms where multiple weak model predictions are aggregated to make the final prediction.

3.2.4 Evaluation Measures and Experimental Setup

The entire experiment was executed on Google Collaboratory, a cloud-based service provided by Google. We used the sci-kit learn package of the Python programming language for initial data prepossessing, feature transformation and classification tasks. The two experimental datasets were divided into training and testing sets through the 10-fold cross-validation technique for model building. Accuracy, ROC, F1-score, precision and recall were considered as evaluation measures to justify the experimental findings. A detailed description of evaluation measures is given in Table 3.4.

3.3 EXPERIMENTAL RESULTS ANALYSIS

The experimental results are briefly discussed in this section.

3.3.1 Analysis of Toddlers Dataset

The performance of different ensemble classifiers on the feature-transformed toddler's dataset is given in Table 3.5. While inspecting the results of the toddler's dataset, AB was found to have the highest accuracy of 99.17% for the MS transformed dataset GBM also achieved 99.17% accuracy on SS transformed dataset, whereas on the RS transformed dataset, AB obtained the best accuracy of 99.25%. In terms of compared

Table 3.5 Performance analysis of toddler dataset.

	FT	AB	GBM	HB	XGB	LM	CB	RF	ET	BG
Accuracy	MS	99.17	98.83	97.59	98.76	98.56	99.04	98.49	98.35	97.11
	SS	99.04	99.17	98.28	98.83	98.35	98.90	98.21	98.28	97.12
	RS	99.25	99.11	98.01	98.90	98.56	98.90	98.21	98.42	96.77
ROC	MS	99.98	99.95	99.94	99.98	99.94	99.99	99.92	99.92	99.34
	SS	99.98	99.94	99.93	99.96	99.92	99.98	99.94	99.92	99.51
	RS	99.98	99.95	99.93	99.97	99.93	99.99	99.92	99.92	99.41
F1-Score	MS	99.1	98.82	98.04	98.81	97.99	98.80	98.19	98.25	96.42
	SS	99.02	98.75	97.97	98.61	97.99	98.72	97.97	98.18	96.75
	RS	99.05	98.74	97.94	98.63	98.45	98.95	98.38	97.99	96.39
Precision	MS	99.9	99.87	99.45	99.87	99.55	99.86	98.60	98.20	98.14
	SS	99.92	99.73	99.57	99.71	99.60	99.75	99.32	98.37	99.03
	RS	99.94	99.56	99.41	99.84	99.29	99.69	99.19	99.01	98.38
Recall	MS	98.31	98.22	96.99	97.31	96.76	97.99	96.53	97.39	94.85
	SS	98.3	98.04	96.85	97.77	96.31	98.15	96.84	98.01	93.73
	RS	98.29	97.85	96.38	97.54	97.15	97.73	97.41	97.11	94.74

the accuracy values on the toddlers, AB was found to have the highest accuracy of 99.25% out of all classifiers for the RS transformed toddler's dataset.

Analyzing the ROC values, the maximum ROC value of 99.99% was obtained by the CB classifier using the MS transformed dataset. Furthermore, the AB and CB classifiers provided the highest ROC value of 99.98% for the SS transformed dataset, whereas on RS transformed dataset, CB obtained best accuracy of 99.99%. Regarding the ROC values, it was seen that the CB classifier delivered the highest ROC values of 99.99% for both the MS and RS transformed toddler's datasets. While inspecting the F1-score on the toddler's dataset, AB was found with the highest F1-score of 99.1%, 99.02% and 99.05% for the MS, SS and RS transformed datasets respectively. In terms of comparing of F1-score values on toddlers, AB was found to have the highest F1-score of 99.1% out of all classifiers for the MS transformed toddler's dataset. While analyzing the experimental precision values it was seen that for all three FT (MS, SS and RS) transformed datasets, highest precision values were obtained by the AB classifier: 99.9%, 99.92% and 99.94%, respectively. Comparing the precision values, it was seen that the AB classifier delivered the highest precision value of 99.94% for the RS transformed toddler's dataset. While inspecting recall values, AB was found to have the highest recall of 98.31%, 98.30% and 98.29% for the MS, SS and RS transformed datasets respectively. In terms of comparing the recall values on toddlers, AB was found with the highest recall of 98.31% out of all classifiers for the MS transformed toddler's dataset.

3.3.2 Analysis of Adults Datasets

The performance of different ensemble classifiers on the feature transformed adults dataset is given in Table 3.6. While investigating the results of the adults dataset, it

Table 3.6 Performance analysis of adult dataset.

	FT	AB	GBM	HB	XGB	LM	CB	RF	ET	BG
Accuracy	MS	98.21	98.54	98.16	98.74	98.06	99.92	98.06	98.54	97.38
	SS	97.89	98.28	98.54	98.54	97.86	98.54	99.92	98.74	96.99
	RS	98.21	98.25	98.45	98.54	98.74	98.74	98.06	97.77	97.18
ROC	MS	99.88	99.87	99.95	99.88	99.95	99.92	99.98	99.98	99.60
	SS	99.91	99.83	99.91	99.93	99.93	99.94	99.94	99.95	99.61
	RS	99.89	99.83	99.94	99.88	99.86	99.97	99.96	99.95	99.40
F1-Score	MS	98.56	98.86	98.27	98.17	98.00	98.65	98.63	98.93	97.38
	SS	98.39	98.83	98.27	98.54	97.85	98.7	98.62	99.06	97.09
	RS	98.74	98.86	98.28	98.56	97.92	98.85	98.50	98.39	96.88
Precision	MS	97.82	97.82	96.44	97.06	96.04	97.50	97.52	98.62	95.32
	SS	97.09	97.01	97.60	96.96	96.55	97.64	97.90	97.99	95.94
	RS	98.02	96.95	96.83	96.97	96.91	98.22	97.79	97.90	96.56
Recall	MS	99.85	99.82	99.09	99.64	99.79	99.59	99.78	99.38	99.39
	SS	99.82	98.58	100	100	99.04	99.85	99.44	99.60	98.05
	RS	99.64	100	99.04	99.60	99.57	99.80	99.62	99.82	99.01

was observed that the highest accuracy of 99.92% was seen from CB and RF using the MS and SS FT methods. As well as reviewing the results of the RS transformed dataset, LM and CB both acquired the same highest results of 98.74%. So, for the adults dataset it was observed from the results analysis that the CB and RF classifiers performed better and delivered 99.92% accuracy for MS and SS transformed datasets respectively. Regarding the ROC results of feature transformed adults datasets, both the RF and ET classifier acquired the highest results of 99.98% on the MS transformed dataset.

Additionally, the highest 99.95% and 99.97% ROC values were seen for the ET and CB classifiers using the SS and RS as FT methods. So, for the adults dataset it was noticed from the results analysis that ET and RF classifiers performed better compared to other classifiers and provided 99.98% ROC using the MS transformed dataset. While investigating the results of adults dataset, it was observed that the highest F1-score result of 98.93% was seen from the ET classifier on the MS transformed dataset. As well as reviewing the results of the SS and RS transformed datasets, ET and GBM achieved the highest F1-score values of 99.06% and 98.86% respectively. So, for the adults dataset it was observed from the results analysis that the ET classifier performed better and obtained a 99.06% F1-score value for the SS transformed dataset.

Regarding the precision results of feature transformed adults datasets, the ET classifier attained the best precision of 98.62% on the MS transformed dataset. Additionally, the highest 97.99% and 98.22% precision values were seen for the RF and CB classifier using the SS and RS as FT methods.

So, for the adults dataset it was noticed from the results analysis that the ET classifier performed better with compared to other classifiers and provided 98.62%

Table 3.7 Risk factors of FT-transformed ASD datasets.

	Toddlers		Adults	
Attribute Name				
	IGAE	CAE	IGAE	CAE
A1_Score	0.16892	0.4058	0.0857	0.22541
A2_Score	0.15251	0.2695	0.0958	0.24587
A3_Score	0.17706	0.3047	0.2456	0.45871
A4_Score	0.29916	0.3062	0.1115	0.36985
A5_Score	0.15161	0.3633	0.6542	0.4584
A6_Score	0.14616	0.2644	0.32569	0.2547
A7_Score	0.26982	0.3652	0.04578	0.2523
A8_Score	0.14363	0.2272	0.02365	0.2546
A9_Score	0.21909	0.4774	0.36455	0.36521
A10_Score	0.002146	0.2549	0.22134	0.14523
Age	0.036194	0.0668	0.00000	0.02583
Gender	0.00526	0.0277	0.00556	0.01452
Ethnicity	0.02584	0.0327	0.08554	0.13658
Jaundice	0.006985	0.0541	0.006874	0.11455
Family member with ASD	0.000131	0.0135	0.0146	0.1144

precision using MS transformed dataset. While investigating the results of the adults dataset, it was observed that highest recall result of 99.85% was seen from the AB classifier on the MS transformed dataset. As well as reviewing the results of the SS and RS transformed datasets, HB, XGB and GBM achieved the highest recall values of 100%. So, for the adults dataset it was observed from the results analysis that the HB, XGB and GBM classifiers performed better and obtained 100% recall value for both the SS and RS transformed datasets.

3.3.3 Discussion

Analyzing the experimental outcomes, the best performing classifiers model predicted ASD with AB (99.25%), RF and CB (99.92%) accuracy; CB (99.99%), RF and ET (99.98%) ROC; AB (99.1%), ET (99.06%) F1-score; AB (99.74%), ET (98.62%) precision; AB (98.31%), GBM, HB, XGB (100.00%) recall; for the toddlers and adults ASD datasets respectively. After analyzing the classification performance of different classifiers, AB for toddlers and RF and CB for adults performed better than the other classifiers for ASD prediction of different stages of life. On the other hand, the classification outcomes of RS transformed datasets for toddlers and the SS transformed datasets for adults showed better performance than any other FT methods. However, we applied two FST approaches, IGAE and CAE, to the RS transformed toddlers and SS transformed adults and calculated the feature importance values for ASD prediction of these datasets which are given in Table 3.7.

Table 3.8 Performance comparison with other studies.

Dataset	Author	Accuracy	ROC	F1 -Score	Precision	Recall
Toddlers	Proposed model	99.25	99.99	99.1	99.74	98.31
Adults	[5]	97.10	-	-	-	-
	[4]	99.85	-	-	-	99.90
	[12]	95.71	-	-	-	85.71
	[16]	96.91	-	-	90.07	96.87
	Proposed Model	99.92	99.98	99.06	98.62	100.00

We also compared our research work with other studies in Table 3.8. A summary of the other studies is provided in Section 3.1. After analyzing Table 3.8, we can observe that our proposed methods outperform previous works.

3.4 CONCLUSION

In order to create prediction models for screening persons of various ages, we acquired two distinct ASD datasets (toddlers, adults) for this study. As a result, we used those ASD datasets to develop three (3) different FT methods (MS, SS, and RS), and then we classed the results using nine (9) different ensemble classification methods (AB, GBM, LM, HB, XGB, CB, RF, ET, BG). In order to determine the top FT techniques and classifier models for these ASD datasets, we also examined the classification results of each FT transformed dataset. To support our experimental findings, we took into account a variety of performance evaluation metrics, including accuracy, ROC, F1-score, precision, recall, MCC, kappa, and log loss. As a result, the doctors can use our suggested prediction models as a substitute or even as a support tool to accurately diagnose ASD instances in patients of various ages. By applying two distinct feature selection strategies, we also investigated the most important traits that were highly predictive for ASD prediction (IGAE, CAE). In that situation, we also assessed each feature's significance for early ASD detection. Our experimental results will therefore aid healthcare professionals in considering the most important factors for efficient decision-making. To diagnose ASD, this study offered four distinct prediction models for four stages of individuals. Our research's shortcoming was that there wasn't enough data to create a universal model that would apply to all life phases. In order to improve the detection of ASD and other neurodevelopmental problems, we plan to gather more data about ASD and build a generalized prediction model for people of all ages.

Bibliography

[1] Allison, C., Auyeung, B., & Baron-Cohen, S. (2012). Toward brief 'Red Flags' for autism screening: The Short Autism Spectrum Quotient and the Short Quantitative Checklist for Autism in toddlers in 1,000 cases and 3,000 controls [corrected]. *J. Am. Acad. Child Adolesc. Psychiatry*, vol. 51, no. 2, pp. 202–212.e7, 2012, doi: 10.1016/J.JAAC.2011.11.003.

[2] F. Thabtah, F. Kamalov, & K. Rajab. (2018). A new computational intelligence approach to detect autistic features for autism screening. *Int. J. Med. Inform.*, vol. 117, pp. 112–124, Sep. 2018, doi: 10.1016/J.IJMEDINF.2018.06.009.

[3] Thabtah, F. & Peebles, D. (2020). A new machine learning model based on induction of rules for autism detection. *Health Informatics J.*, vol. 26, no. 1, pp. 264–286, Mar. 2020, doi: 10.1177/1460458218824711.

[4] K. S. Oma, P. Mondal, N. S. Khan, M. R. K. Rizvi, & M. N. Islam. (2019). A machine learning approach to predict autism spectrum disorder. *2nd Int. Conf. Electr. Comput. Commun. Eng. ECCE 2019, Apr. 2019.* doi: 10.1109/ECACE.2019.8679454.

[5] Abbas, H., Garberson, F., Glover, E., & Wall, D.P. (2018). Machine learning approach for early detection of autism by combining questionnaire and home video screening. J. Am. Med. Inform. Assoc., vol. 25, no. 8, pp. 1000–1007, Aug. 2018. doi: 10.1093/JAMIA/OCY039.

[6] Thabtah, F. (2019). Machine learning in autistic spectrum disorder behavioral research: A review and ways forward. *Inform. Health Soc. Care*, vol. 44, no. 3, pp. 278–297, Jul. 2019, doi: 10.1080/17538157.2017.1399132.

[7] K. L. Goh, S. Morris, S. Rosalie, C. Foster, T. Falkmer, & T. Tan. (2016). Typically developed adults and adults with autism spectrum disorder classification using centre of pressure measurements. *ICASSP, IEEE Int. Conf. Acoust. Speech Signal Process. - Proc.* vol. 2016-May, pp. 844–848, May 2016, doi: 10.1109/ICASSP.2016.7471794.

[8] Crippa, A. et al. (2015). Use of machine learning to identify children with autism and their motor abnormalities. *J. Autism Dev. Disord.*, vol. 45, no. 7, pp. 2146–2156, Jul. 2015, doi: 10.1007/S10803-015-2379-8.

[9] B. Tyagi, R. Mishra, & N. Bajpai. (2018). Machine learning techniques to predict autism spectrum disorder. *1st Int. Conf. Data Sci. Anal. PuneCon 2018 - Proc.* Nov. 2018, doi: 10.1109/PUNECON.2018.8745405.

[10] Duda, M., Ma, R., Haber, N., & Wall, D.P. (2016). Use of machine learning for behavioral distinction of autism and ADHD. *Transl. Psychiatry*, vol. 6, no. 2, 2016, doi: 10.1038/TP.2015.221.

[11] S. B. Shuvo, J. Ghosh, & A. S. Oyshi. (2019). A data mining based approach to predict autism spectrum disorder considering behavioral attributes. *2019 10th Int. Conf. Comput. Commun. Netw. Technol. ICCCNT 2019.* Jul. 2019, doi: 10.1109/ICCCNT45670.2019.8944905.

[12] O. Altay & M. Ulas. (2018). Prediction of the autism spectrum disorder diagnosis with linear discriminant analysis classifier and K-nearest neighbor in children. *6th Int. Symp. Digit. Forensic Secur. ISDFS 2018 - Proceeding.* vol. 2018-January, pp. 1 – 4, May 2018, doi: 10.1109/ISDFS.2018.8355354.

[13] F. N. Buyukoflaz & A. Ozturk. (2018). Early autism diagnosis of children with machine learning algorithms. *26th IEEE Signal Process. Commun. Appl. Conf. SIU 2018*. pp. 1 –4, Jul. 2018, doi: 10.1109/SIU.2018.8404223.

[14] M. F. Misman et al.. (2019). Classification of adults with autism spectrum disorder using deep neural network. *Proc. - 2019 1st Int. Conf. Artif. Intell. Data Sci. AiDAS 2019*. pp. 29–34, Sep. 2019, doi: 10.1109/AIDAS47888.2019.8970823.

[15] H. Talabani & E. Avci. (2019). Performance Comparison of SVM Kernel Types on Child Autism Disease Database. *2018 Int. Conf. Artif. Intell. Data Process. IDAP 2018, Jan. 2019*. doi: 10.1109/IDAP.2018.8620924.

[16] UCI Machine Learning Repository: Autism Screening Adult Data Set. https://archive.ics.uci.edu/ml/datasets/Autism+Screening+Adult (accessed Jul. 14, 2021).

[17] Autism screening data for toddlers — Kaggle. https://www.kaggle.com/fabdelja/autism-screening-for-toddlers (accessed Jul. 14, 2021).

CHAPTER 4

Implementing Machine Learning Using the Neural Network for the Time Delay SIR Epidemic Model for Future Forecasting

Sayed Allamah Iqbal

Department of Electrical & Electronic Engineering, International Islamic University Chittagong, Chattogram, Bangladesh

Md. Golam Hafez

Department of Mathematics, Chittagong University of Engineering and Technology, Chattogram, Bangladesh

A.N.M. Rezaul Karim

Department of Computer Science & Engineering, International Islamic University Chittagong, Chattogram, Bangladesh

CONTENTS

FORECASTING is significant for future stability. The dynamic analysis provides the characteristics and behavior of various kinds of physical phenomena. Nevertheless, in some dynamic systems, which are based on unknown parameters, it isn't easy to obtain a model function to predict the future. On the flip side, machine learning algorithms are based on data-driven solutions and are not essential mathematical models for foresight. This work forecasts the future of contagious diseases by considering the time delay SIR infectious mathematical model. The data-driven dynamic analysis is made for this model by applying Neural Network (NN) techniques. This function-free data-driven analysis is examined for future predictions of infectious diseases. Notably, these exertions observe that the function-free forecast is similar to predicting the time delay nonlinear SIR model function.

4.1 INTRODUCTION

In the numerous fundamental physical problems in science and industrial fields, nonlinear dynamics are significant in analyzing their complex phenomena. For instance, epidemiology, sociology, neuroscience, and physics are widely defined mathematical models of complex systems that incorporate nonlinear dynamics. Along with this theoretical approach of the physical problems in mathematics, science, and engineering, the involvement of machine learning is furnishing a distinct dimension for the analysis of these model equations.

Nowadays, many branches of engineering, physical science, biological and chemical science are evolving the data-driven modeling of complex problems [2]. These complex problems are difficult to model, like a mathematical function, due to their undiscovered function parameters. For instance, financial markets, turbulent fluids, network neurons, climate change, or epidemiology are nonlinear dynamical systems, which display multiscale phenomena in both space and time. The data-driven modeling in these field is an excellent example for deep neural networks in machine learning. Apply the data-driven [1] dynamical model forecast in the field of engineering, natural science, or physical science. It is possible to obtain algorithms and innovation in machine learning to get the significant space-time related pattern for future prediction [3] of dynamical activity. This data-driven activity of the dynamical system is essential for understanding the scientific process, which will enhance the capability of state estimation and give control of the complex system. Although it is significant to develop our knowledge of complex problems, such as epidemic disease, the mathematical models are related to some known parameters such as the service factor, disease-related factor, sociological factor, environmental factor, and some other unpredictable factors. But there is an ample amount of data available for these problems, whereas models frequently remain obscure.

Nobel laureates Hubel and Wiesel [6], won for research work on the primary visual cortex of a cat experiments. It showed that neuronal networks (NNs) were organized in hierarchical cell layers for visual [10] processing of stimuli. They are the impulse of NNs in machine learning.

In 1980, Fukushima [4], first introduced the mathematical [9] model of NNs in pattern recognition. Currently numerous characteristic traits of NNs in the multilayer structure, convolution, and max pooling are well-known applied method for the analysis of data-driven nonlinear dynamical models, and many more physical problems.

In this work, the effectiveness of the NN technique has been exemplified in the time-delay SIR model.In the early 20th-century, Kermack and McKendrickr [9] developed the well-known SIR mathematical model for infectious diseases, which includes three sections—susceptible (S), infectious (I), and recovered (R). This popular SIR model is:

$$\begin{aligned}\frac{dS(\eta)}{d\eta} &= \gamma - \beta S(\eta)I(\eta) - \mu S(\eta),\\ \frac{dI(\eta)}{d\eta} &= \beta S(\eta)I(\eta) - rI(\eta) - \mu I(\eta),\\ \frac{dR(\eta)}{d\eta} &= rI(\eta) - \mu R(\eta),\\ N(\eta) &= S(\eta) + I(\eta) + R(\eta), \text{ where } N \text{ is the total population.}\end{aligned} \tag{4.1}$$

The model parameters $S(\eta)$, $I(\eta)$, and $R(\eta)$ denote the number of people in the population susceptible to the disease, the number of infectious people with some virus of time η, and the number of people in the population who have been removed from the system through temporal immunity, respectively.

The flow of this research progress with an introduction is supported by the time delay SIR epidemic model; subsequently, the neural networks for the time-delay SIR model are finally furnished with discussion and conclusion.

4.2 TIME DELAY SIR EPIDEMIC MODEL

Apply the distributive [14, 15] time delay in Eq. (4.1), which developed the following SIR model equation of time delay;

$$\begin{aligned}\frac{dS(\eta)}{d\eta} &= \gamma - \beta I(\eta)\int_{-\infty}^{t}\phi(\eta-\tau)d\tau S(\eta) - \mu S(\eta),\\ \frac{dI(\eta)}{d\eta} &= \beta I(\eta)\int_{-\infty}^{t}\phi(\eta-\tau)d\tau S(\eta) - rI(\eta) - \mu I(\eta),\\ \frac{dR(\eta)}{d\eta} &= rI(\eta) - \mu R(\eta),\\ \text{where, } \phi(\eta) &= \alpha e^{-\alpha\eta}, \ \& \int_{0}^{\infty}\phi(\eta)d\tau = 1.\end{aligned} \tag{4.2}$$

The variable $R(\eta)$ is absent in the first and second equations of Eq. (4.2), so considering the variable $S(\eta)$ and $I(\eta)$, the following modified model is:

$$\begin{aligned}\frac{dS(\eta)}{d\eta} &= \gamma - \beta I(\eta)\int_{-\infty}^{t}\phi(\eta-\tau)d\tau S(\eta) - \mu S(\eta),\\ \frac{dI(\eta)}{d\eta} &= \beta I(\eta)\int_{-\infty}^{t}\phi(\eta-\tau)d\tau S(\eta) - rI(\eta) - \mu I(\eta).\end{aligned} \tag{4.3}$$

Apply the chain rule formula in Eq. (4.3) by choosing the

$$V(\eta) = \int_{-\infty}^{t}\phi(\eta-\tau)d\tau.$$

The final converted distributed delay of the SIR model is:

$$\begin{aligned}\dot{S} &= \gamma - \beta I(\eta)z(\eta) - \mu S(\eta),\\ \dot{I} &= \beta I(\eta)V(\eta) - rI(\eta) - \mu I(\eta),\\ \dot{V} &= \alpha\left(S(\eta) - V(\eta)\right).\end{aligned} \tag{4.4}$$

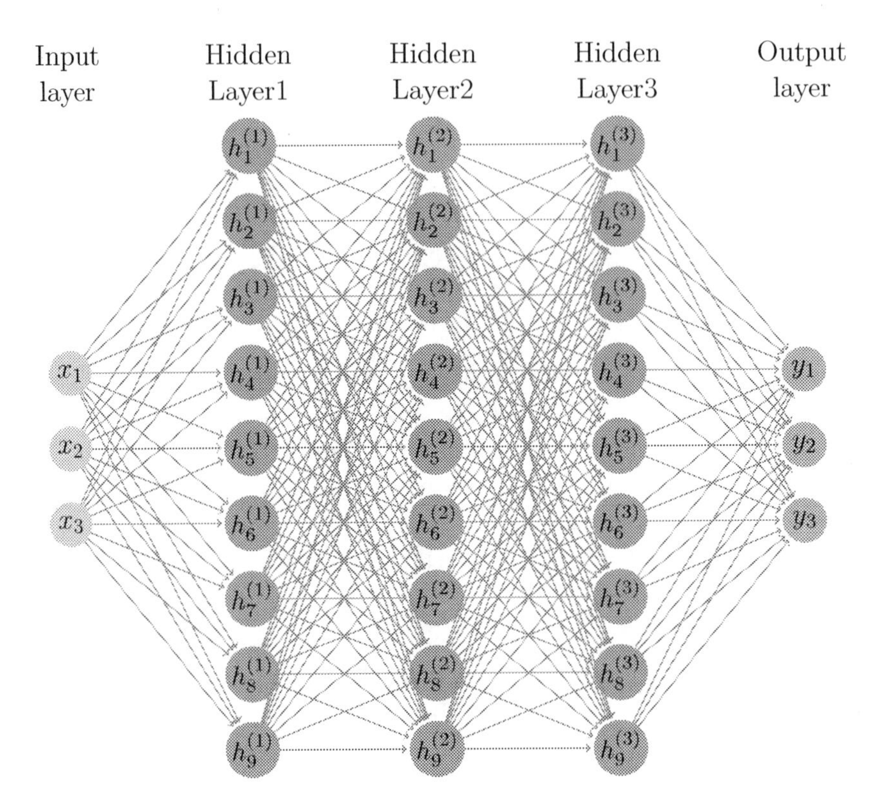

Figure 4.1 Illustration of time delay SIR neural net architecture mapping an input for a random initial state point layer (x_1, x_2, x_3) to an output layer (y_1, y_2, y_3) across the three hidden layers. Each hidden layer consists of nine neurons.

4.2.1 Neural Networks for time-delay SIR model

A distinct set of mathematical tasks is now employing neural networks. It is a fantastic architecture to examine function-free mathematical model future prediction. In this effort, supervised learning is applied in state prediction of the data-driven [10, 11], time-delay SIR model through the implication of NNs in Eq. (4.4).

Model Eq. (4.4) is reset by substituting the variables $S(\eta) = x_1(\eta), I(\eta) = x_2(\eta)$, $V(\eta) = x_3(\eta)$:

$$\begin{aligned} \dot{x_1} &= \gamma - \beta y(\eta) x_3(\eta) - \mu x_1(\eta), \\ \dot{x_2} &= \beta x_2(\eta) x_3(\eta) - r x_2(\eta) - \mu x_2(\eta), \\ \dot{x_3} &= \alpha \left(x_1(\eta) - x_3(\eta) \right). \end{aligned} \tag{4.5}$$

The state of Eq. (4.5) is picked as a computer-generated random input layer $\{x_1, x_2, x_3\}$. The parameter values from Wencai Zhao [15] $\gamma = 1.8, \beta = 0.08, \mu = 0.2, r = 0.6$, and $\alpha = 0.1$ for generating a corresponding output layer $\{y_1, y_2, y_3\}$. Three hidden NN layers, individually containing nine neurons, are depicted in Fig. 4.1.

This work describes how to train the NNs to learn to update the rule that advances the phase space from x_i to x_{i+1}, where i is the phase state of the system at time η_i.

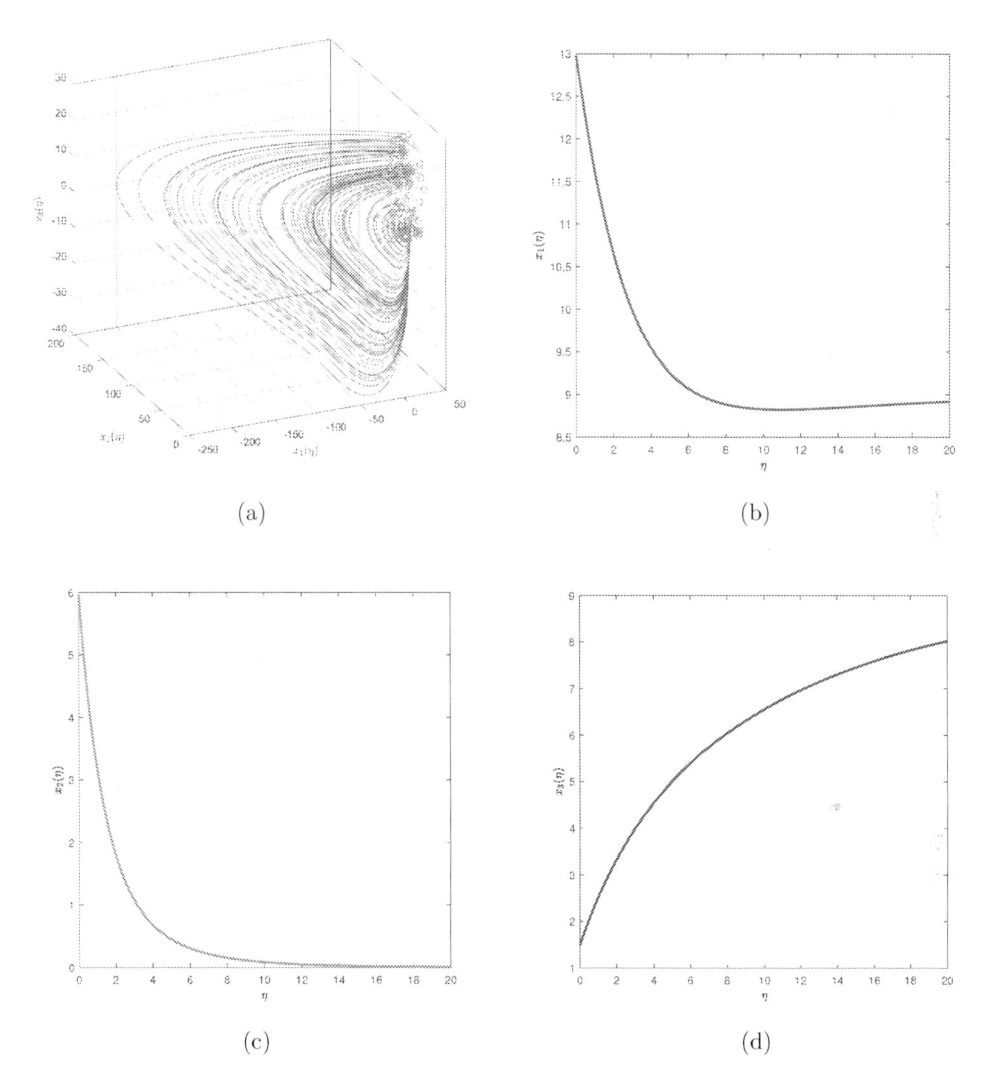

Figure 4.2 Development of the SIR time-delay dynamical equations (4.5) for one hundred randomly chosen initial states (red circles). For the parameters $\gamma = 1.8, \beta = 0.08, \mu = 0.2, r = 0.6$, and $\alpha = 0.1$, all trajectories conform to a stable point at $(6.25, 1.1, 6.25)$. These trajectories, created from a diverse set of initial data, train NNs to learn the nonlinear mapping from x_i to x_{i+1}. Subsequently, the connected time series plot for the susceptible (x_1) in (b), infected (x_2) in (c), and recovery (x_3) in (d) are defined the stable time-delay SIR model.

Precisely solving in time needs a nonlinear transfer function because the time-delay SIR itself is nonlinear. The following NNs trained trajectories to advance the solution at $\delta t = 0.01$ into the future prediction. The simulation figures for Eq. (4.5) are generated for 100 initial random phase state values. The function-free data-driven prediction plot is created for two random phase states for 2 and 22 random initial phase states, which are depicted in Fig. 4.2 and Fig. 4.3.

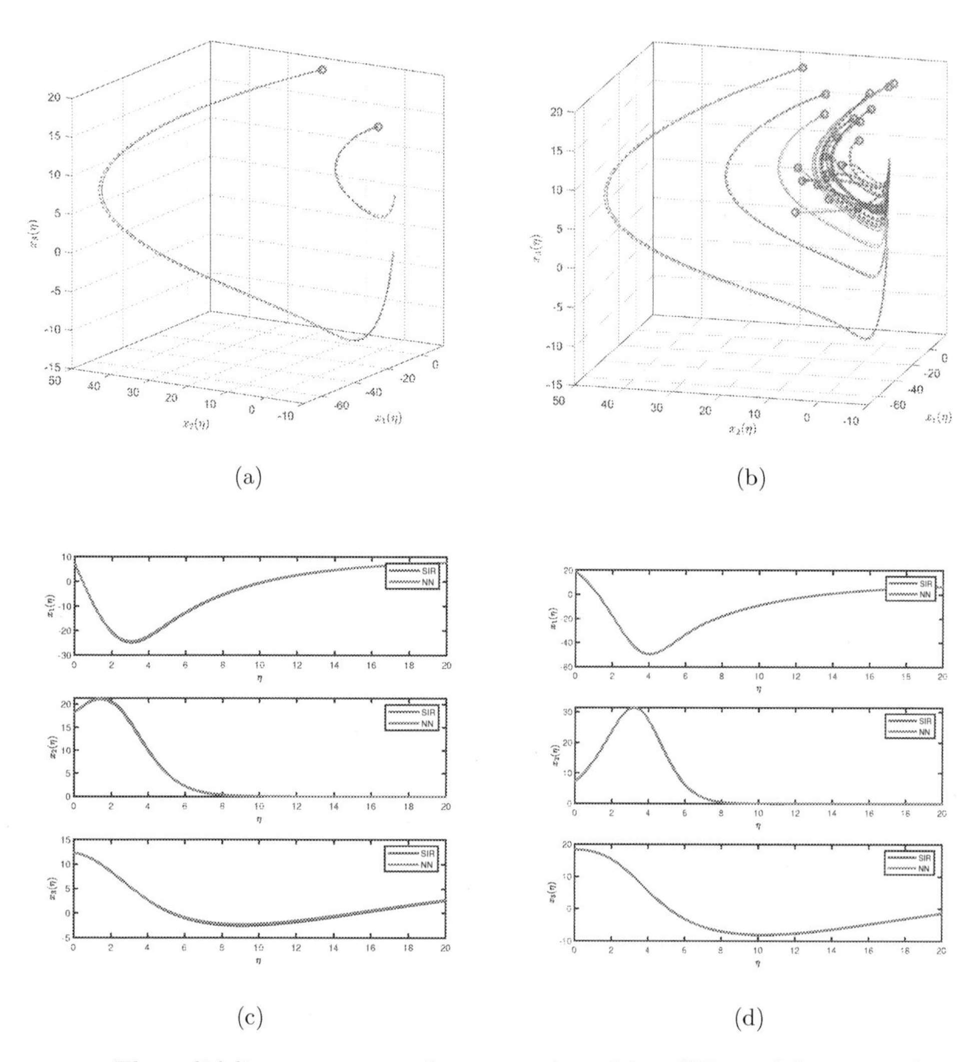

Figure 4.3 The solid lines represent the exact time-delay SIR model testing data solution for the time-delay trajectories. The dotted lines represent the prediction trajectories of NNs in (a) represent two(2) trained NNs and (b) displaying twenty-two (22) forecast trained NNs, and feature distribution of training/testing data for components x_1, x_2, x_3 regarding time(η). With two training NNs, trajectory time series in (c) for the one simulation. (d) Second, simulation trajectory time series express trained data(NNs) and exact solution testing data for the model.

4.3 DISCUSSION

High-dimension interpolation problems in supervised machine learning play a significant role in the data-driven dynamic system. Accurate pattern recognition is for future prediction [12, 13], of the nonlinear dynamic system; the hidden layer of NNs is important since less or more neurons in hidden layers create the drawback of under-fitted or over-fitted patterns. Moreover, the simulation considers the hidden layers by selecting the arbitrary neurons in their respective layers to get a satisfactory outcome of the SIR time-delay model.

The computer simulation generates one hundred random initial conditions in the classified red dots to trace one hundred trajectories using the adaptation 4th-order Runge-Kutta method [7], represented in Fig. 4.2. In contrast, Fig. 4.2 illustrates a stable dynamic system with corresponding time series concerning the susceptible (x_1), infected (x_2), and recovered (x_3) populations about time η.

The NN simulation produces from time $\eta = 0$ to $\eta = 20$, with time step $\delta\eta = 0.01$ for the net of the SIR function Eq. (4.5). These are used as a new dynamical driven data set prediction of the future illustrated from Fig. 4.2 to 4.4. Especially Fig. 4.2 displays the exact dynamic phase state [7, 8, 13], evaluation of a 4th-order Runge-Kutta numerical solution. Subsequently, Fig. 4.2 also displays the time series analysis. Then, randomly taking initial state(x_1, x_2, x_3) for the input layer, the NNs are trained on the trajectory data, generate the nonlinear mapping [5], and x_i to x_{i+1} are used for the future forecasting of the SIR model. It is compared with the 4th-order Runge-Kutta solution of the SIR model Eq. (4.5) with Fig. 4.2 to Fig. 4.3 for twenty unit times in Fig. 4.3(a) two arbitrary red dot initial states give the same trajectory compared with the solid lines to dotted lines of Fig. 4.3a. In contrast, the same amount of twenty unit time with the same $\delta\eta = 0.01$, for 22 random initial states gives the same trajectories compare with solid lines and the dotted lines NNs in 4.3b. Subsequently, the time series analysis of the numerical solutions compared with the NN data simulation is displayed in Fig. 4.3(c) and Fig. 4.3(d).

Fig. 4.4(a) shows the network structure along with the performance of the training over 1000 epochs of training. The NN converges firmly to a network that creates efficiency on the best validation performance order at 7.2334×10^{-07} at epoch 1000. The cross-validation results are also shown in the training state, the error histogram, and regression analysis in Figs. 4.4(b) to 4.4(d).

4.4 SUMMARY

Covid-19, SARS, and avian influenza-like other infectious diseases are a tremendous continuous threat for human beings. This data-driven nonlinear function-free NN is helpful in predicting Covid-19-like infectious diseases.

The NN technique's effectiveness is verified for different benchmark problems, for instance, the Lorenz nonlinear chaotic dynamic system and some other complex system forecasts and identification.

Besides, the NN technique for data-driven dynamic analysis for the future state forecast has a deficiency of long-term temporal dependencies from dynamic time-series data.

This work mainly addresses theoretical rather than experimental data for forecasting epidemic infectious diseases. Computer simulations use 100 random initial guesses to locate probabilistic outcomes rather than deterministic ones. The comparison between the simulation of the SIR mathematical model equation contrasts the simulated initial guess of the random data-driven function-free model by applying the NN gives the remarkable similarity of both simulations. Future research can be

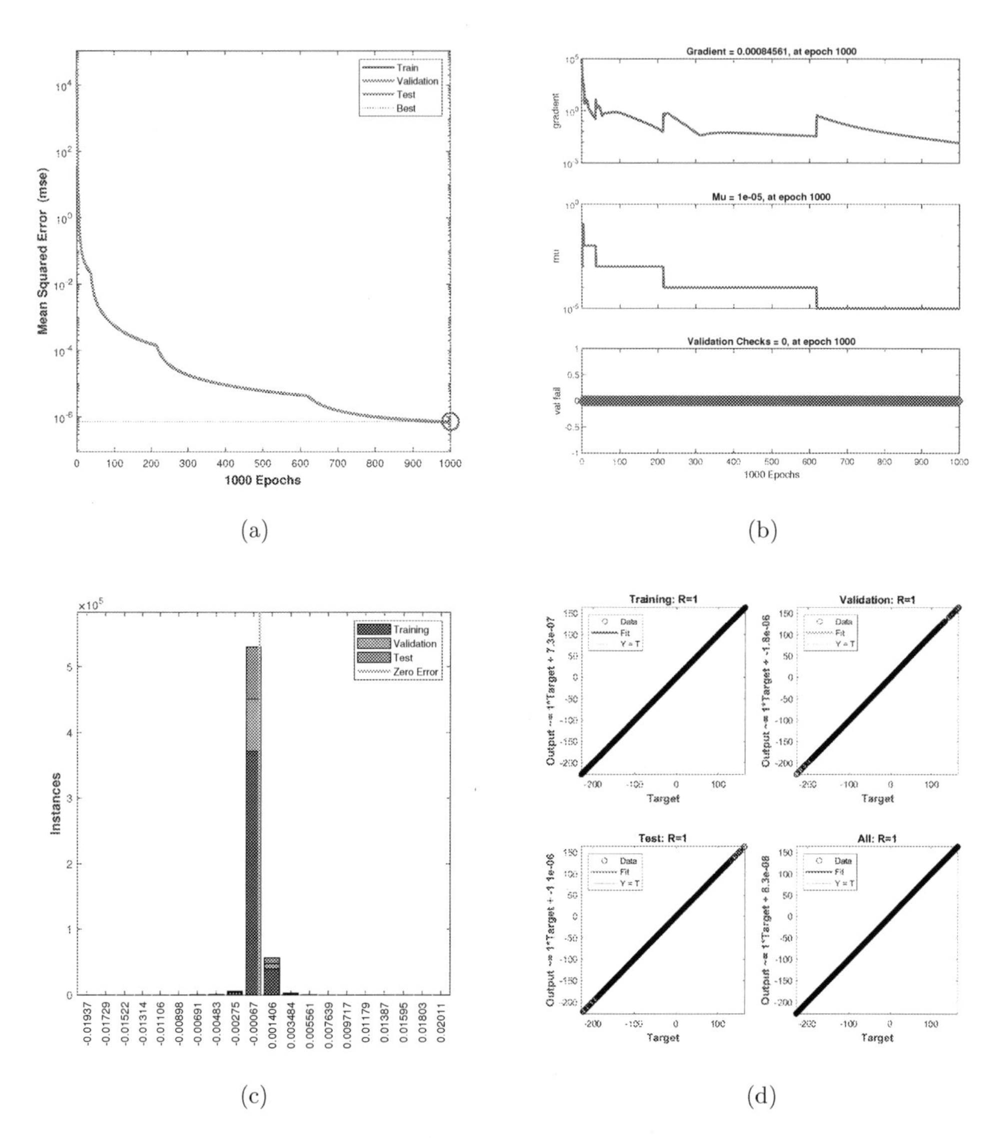

Figure 4.4 Illustration in (a) for the training of the NNs over the number of epochs, in (b) display the training state of NNs, after that in (c) the error performance of the NNs architecture for training, validation and test, and in (d) the NNs data regression analysis of the future prediction of the SIR model.

done by connecting experimental data to perceive the elusive knowledge of future forecasting for epidemic diseases like Covid-19.

ACKNOWLEDGEMENT(S)

We acknowledge and are grateful to the anonymous reviewers for their detailed direction to improve the research output.

DISCLOSURE STATEMENT

The authors of this article declare that they have no conflicts of interest. This article used artificial computer-created random data and did not contain any actual data for epidemic diseases, like Covid-19.

Bibliography

[1] Brunton, S., J. Proctor, & J. Kutz. (2016). Discovering Governing Equations from Data by Sparse Identification of Nonlinear Dynamical Systems. *Proceedings of the National Academy of Sciences*. https://doi.org/10.1073/pnas.1517384113.

[2] Fujii, Keisuke, & Yoshinobu Kawahara. (2019). Dynamic Mode Decomposition in Vector-Valued Reproducing Kernel Hilbert Spaces for Extracting Dynamical Structure among Observables. *Neural Networks* (117) (September): 94–103. https://doi.org/10.1016/j.neunet.2019.04.020.

[3] Fujii, Keisuke, Takeshi Kawasaki, Yuki Inaba, & Yoshinobu Kawahara. (2018). Prediction and Classification in Equation-Free Collective Motion Dynamics. *PLOS Computational Biology* 14 (11): e1006545. https://doi.org/10.1371/journal.pcbi.1006545.

[4] Fukushima, K. (1980). Neocognitron: A Self Organizing Neural Network Model for a Mechanism of Pattern Recognition Unaffected by Shift in Position. *Biological Cybernetics* 36 (4): 193–202. https://doi.org/10.1007/BF00344251.

[5] Hafez, M.G., S.A. Iqbal, S. Akther, & M.F. Uddin. (2019). Oblique Plane Waves with Bifurcation Behaviors and Chaotic Motion for Resonant Nonlinear Schrodinger Equations Having Fractional Temporal Evolution.*Results in Physics* 15 (December): 102778. https://doi.org/10.1016/j.rinp.2019.102778.

[6] Hubel,D.H., & Wiesel,T.N.(1962).Receptive Fields, Binocular Interaction and Functional Architecture in the Cat's Visual Cortex. *The Journal of Physiology* 160 (1): 106–54. https://doi.org/10.1113/jphysiol.1962.sp006837.

[7] Iqbal, S.A., M.G. Hafez, & Samsul Ariffin Abdul Karim. (2020). Bifurcation Analysis with Chaotic Motion of Oblique Plane Wave for Describing a Discrete Nonlinear Electrical Transmission Line with Conformable Derivative." *Results in Physics* 18 (September): 103309. https://doi.org/10.1016/j.rinp.2020.103309.

[8] Iqbal, S.A., & M.G. Hafez. (2020). Dynamical Analysis of Nonlinear Electrical Transmission Line through Fractional Derivative. In 2020 23rd International Conference on Computer and Information Technology (ICCIT), 1–5. https://doi.org/10.1109/ICCIT51783.2020.9392656.

[9] Kermack, W.O., & Mckendrick, Á. (1927). A Contribution to the Mathematical Theory of Epidemics. https://doi.org/10.1098/RSPA.1927.0118.

[10] Mallat, Stéphane. (2016). Understanding Deep Convolutional Networks. *Philosophical Transactions of the Royal Society A: Mathematical, Physical and Engineering Sciences* 374 (2065): 20150203. https://doi.org/10.1098/rsta.2015.0203.

[11] Mangan, N., J. Kutz, S. Brunton, & J. Proctor. (2017). Model Selection for Dynamical Systems via Sparse Regression and Information Criteria. Proceedings of the Royal Society A: Mathematical, Physical and Engineering Sciences. https://doi.org/10.1098/rspa.2017.0009.

[12] Tanaskovic, Marko, L. Fagiano, C. Novara, & M. Morari. (2017). Data-Driven Control of Nonlinear Systems: An On-Line Direct Approach. *Autom.* https://doi.org/10.1016/j.automatica.2016.09.032.

[13] Teng, Qi, & Lei Zhang. (2019). Data Driven Nonlinear Dynamical Systems Identification Using Multi-Step CLDNN. *AIP Advances* 9 (8): 085311. https://doi.org/10.1063/1.5100558.

[14] Yan, Ping, & Shengqiang Liu. (2006). SEIR Epidemic Model with Delay. *The ANZIAM Journal* 48 (1): 119–34. https://doi.org/10.1017/S144618110000345X.

[15] Zhao, Wencai,Tongqian, Z., Zhengbo, C., Xinzhu, M., & Yulin, L. (2013). Dynamical Analysis of SIR Epidemic Models with Distributed Delay. *Journal of Applied Mathematics* 2013 (July): e154387. https://doi.org/10.1155/2013/154387.

CHAPTER 5

Prediction of PCOS Using Machine Learning and Deep Learning Algorithms

Syed Mohd. Farhan

Department of Computer Science & Engineering, East Delta University, Chattogram, Bangladesh

Maimuna Manita Hoque

Department of Computer Science & Engineering, East Delta University, Chattogram, Bangladesh

Mohammed Nazim Uddin

School of Science, Engineering & Technology, East Delta University, Chattogram, Bangladesh

CONTENTS

In recent times, the whole population is facing severe medical conditions, most of which affect women. Infertility, abortions, anovulation etc. are some common medical issues found among women. Polycystic Ovary Syndrome (PCOS) is a condition where ovarian cysts grow due to eggs not being able to release over time and the follicles that keep growing inside create multiple cysts. Due to the variety of symptoms associated with, and numbers of gynecological disorders, it is difficult to diagnose PCOS. Moreover, the symptoms associated with PCOS can be the symptoms for other diseases as well. Due to the complex process in diagnosing PCOS, this chapter is aimed towards building a model to help physicians in detecting PCOS. During consultations with doctors and clinical trials in Kerala, India, this chapter investigated different algorithm techniques and found effective techniques used in a data set of 541 women. PCOS classification is done through many machine learning techniques such as the Logistic Regression classifier, Naive Bayes classifier, AdaBoost classifier, Random Forest Classifier and Artificial Neural Networks from deep learning with the help of Google Colaboratory. After feature engineering and hyperparameter tuning, the logistic regression classifier showed the most accuracy for PCOS prediction with 95.40% and has a recall score of 94.12%.

5.1 INTRODUCTION

Many patients have been saved in recent years by technical advancements in healthcare and we humans are continually improving our quality of life. Not only this, but in the medical field, technology has had huge impacts on almost every medical practice. Originally, artificial intelligence was aimed at making machines more useful and capable of independent thought. Machine learning is now helping to create automation in developing analytical models.

Polycystic ovary syndrome is among the most common hormonal disorders seen in women of childbearing age (PCOS) [1]. This is a heterogeneous endocrine syndrome which is especially prone to anovulation, miscarriage, heart disease, type 2 diabetes, obesity, etc. A variety of symptoms or complications are encountered by women with PCOS, including gynecological disorders such as ovulation failure, endometrial cancer, late menopause and infertility, cardiac disorders such as cardiovascular disease and hypertension, metabolic disorders such as dyslipidemia, insulin resistance, type 2 diabetes, physical disorders such as hair loss and baldness, hirsutism, central obesity, acne, and psychological disorders [2]. The prevalence of PCOS is extremely variable, ranging from 2.2 percent to 26 percent globally. Prevalence figures in China range from 2 percent to 7.5 percent and 6.3 percent in Srilanka. The prevalence of PCOS (by Rotterdam criteria) was 9.13 percent and 22.5 percent, respectively (10.7 percent by Androgen Excess Society criteria), in studies performed in South India and Maharashtra [3]. Early diagnosis of PCOS is critical since an increased risk of developing many medical problems has been associated with it including heart disease, high cholesterol, insulin resistance, high blood pressure and type 2 diabetes. As detecting PCOS is tricky, a large number of tests are needed to be done to figure out the symptoms of the patient as doctors have to look at many factors. This chapter proposes a roadmap to construct a model by applying computer intelligence to predict those prone to Polycystic Ovary Syndrome (PCOS) and visualize the PCOS probability. Using supervised machine learning methods, through analyzing various features and their differences between PCOS-affected patients and non-PCOS-affected patients, this chapter aims to build a model that can predict whether a patient has PCOS or not. This chapter focuses on classifying non-PCOS patients and PCOS-affected patients by using several supervised machine learning methods and evaluating their performance to take into account which method works best for this work.

The key components of the proposed predictive model for PCOS are:

a. constructing a reference dataset of clinical information from health care practitioners for patients with and without PCOS,

b. selecting the most impactful features applicable to the solution,

c. constructing a predictive model (using a machine learning technique) using the sample as a dataset of preparation, and

d. evaluating the accuracy of the PCOS diagnosis.

5.2 RELATED WORK

Gulam Saidunnisa Begum et al. have investigated the general factors of PCOS such as heredity, fast food consumption habits, participation in physical exercise, waist size and body mass index of study participants, which were taken as possible risk factors for developing PCOS with the information of family history of PCOS [4]. This research concluded that patients with family history of PCOS have higher risk for developing PCOS.

Amsy Denny et al. proposed a system for detecting and predicting PCOS, using various clinical and metabolic features [5]. This study identified 8 features out of 23 clinical and metabolic features. The study conducted in Thrissur inferred that biochemical profile or the product of USG alone could not be selected as a PCOS treatment diagnostic method as both PCOS and infertility-related variables fall into both groups. AMH is found to be a very promising feature for detecting PCOS and infertility according to their findings.

Ninety-seven percent is the highest accuracy obtained in [6], 82 percent in [7], 93.9 percent in [8] and 90 percent in [9], contrasting the accuracies obtained for other PCOS detection tests.

Sharvari S. Deshpande et al. proposed automated detection of PCOS where ovarian ultrasound images are used to calculate follicle numbers and some biochemical, clinical and imaging criteria are taken into account for classifying patients as PCOS affected and normal [10].

Palak Mehrotra and her co-authors proposed an algorithm that involves clinical and metabolic features for selecting feature vectors [8]. In Kolkata, this study of female patients found that clinical and metabolic characteristics were needed for formulating the function vector of their algorithm. Using a two-sample t-test, the features which are significantly responsible for discrimination between normal and PCOS groups are selected. Bayesian and Logistic Regression classifiers were used in this study, and the Bayesian classifier performed better with 93.93 percent accuracy than the logistic regression's 91.04 percent accuracy. Bayesian classifier also gave better precision. Its study was conducted with a relatively large amount of data. But again, the paper is aimed more towards comparison of classifiers rather than preparing a proper framework for proper diagnosis of PCOS.

Cheng et al. assessed the results of a rule-based classification and a gradient boosted model in pelvic ultrasounds and proposed a model for the automated extraction of features and classification of polycystic ovary morphology (PCOM) [6].

When considering the relationship between PCOS and the infertility rate in women reporting PCOS, Stepto et al. found the use of fertility hormone therapy was substantially higher [11]. Due to the prevalence of PCOS and the health and economic impacts of infertility, strategies for improving PCOS diagnoses and contributing factors for fertility are important.

B. Cahyono et al. approached deep learning in order to classify ultrasound images. A CNN (Convolutional Neural Network) was used for the extraction of features [12]. The study suggested that optimizing CNN is very difficult. The author proposed a higher dropout rate in future work.

5.3 METHODOLOGY

The proposed system tries to predict if a patient is affected with PCOS or not and also visualizes the prediction probability. Various machine learning techniques are used where recall, accuracy, precision and F1 score are evaluated, and based on evaluation metrics comparison, a final model is then selected and deployed in the PCOS Predictor prototype for visualization purposes. The system architecture for the proposed system is given in Figure 5.1.

Machine learning allows the user to feed a vast amount of data to a computer algorithm and have that computer analyze and make suggestions and decisions based on the input data only. The machine learning model will determine whether or not a patient has PCOS by analyzing the data of patients on the basis of their various features.

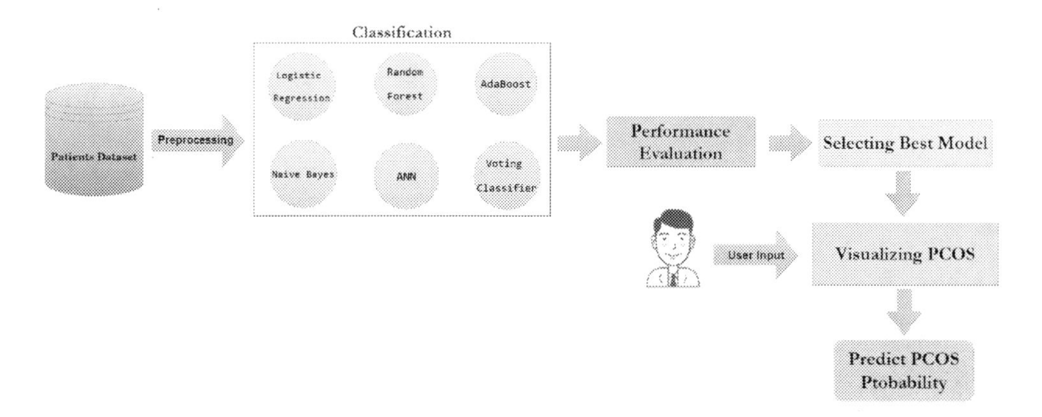

Figure 5.1 Proposed System Architecture

5.3.1 Dataset Collection

The dataset was collected from Kaggle, which is world's biggest community for data science. It was uploaded by Prasoon Kottarathil, a data scientist of Muriyad, Kerala, India. The dataset has two files containing 541 samples of patients with infertility and without infertility. These samples are of patients from 10 hospitals and clinics in the adjacent area of Thrissur, Kerala, India. Patients gave the information with signed consent which promised complete anonymity. Patients were given a proforma, which was used for obtaining this information. Later, the collected data was consolidated into a single sheet of data. There are 42 features, which were obtained from scan findings, blood tests, medical history, different kinds of symptoms of patients and hormone analysis.

5.3.2 Data Preprocessing

There are several steps in data pre-processing for this proposed system. For further analysis of the data, the dataset needed to be cleaned to make it more error-free and suitable for applying machine learning techniques.

5.3.2.1 Data Cleaning

A dataset can contain many irrelevant and incorrect information. This part of the architecture handles corrupted, incomplete, incorrect, duplicate or missing data where the missing values are assigned the median value of the feature. Also, outliers, which can simply be human input errors, have to be identified for each feature and then removed so that the algorithms can learn better in the training phase. Domain knowledge of some features are required to handle the extreme values as some rare cases can be possible for patients. Redundant information is also removed.

5.3.2.2 Feature Engineering

Categorical Feature was used to generate new numerical features. Also features like BMI, Waist: Hip Ratio, Follicle Numbers of the right and left ovaries can help generate new features, which could help the model find new relations from these features. One hot encoding and binning method is used for creating new features from existing features.

5.3.2.3 Feature Selection

One of the most important steps of this architecture involves selecting features which can help the algorithms by removing irrelevant noisy information that could cause them to classify incorrectly, resulting in overfitting or underfitting. The filter method is used for feature selection and filters out the important features as a subset of all features. The Pearson correlation matrix is used for finding the correlation of all the features. The correlation coefficient values range from -1 to 1 where -1 represents negative linear correlation, 0 implies no linear correlation and 1 represents positive linear correlation. Features having correlation values above 0.25 were selected. As the filter method is faster and can detect highly correlated features using less storage, it is used to drop various highly correlated features.

5.3.2.4 Feature Scaling

Z-score normalization or standardization is done as a part of feature scaling to generate a fixed range among independent features which can handle variance among the values of these features. It is a feature scaling technique used to rescale the features ensuring the mean of the feature is 0 and standard deviation is 1.

5.3.3 Dataset Split

At first, the dataframe is split with 1:4 ratio into 20% validation data and 80% training-testing data. After that, the Scikit-Learn library is used for splitting the training testing dataframe into 20% testing data and 80% training data with 1:4 ratio, again using a simple random sampling technique. The testing set is used for evaluating the performance of the models and the training set is used to train the models with a subset of the dataset. The validation set is unseen data for the models which is used on the final model. It is used for validating the final model.

5.3.3.1 Handling Imbalanced Data

SMOTE, or the Synthetic Minority Oversampling Technique, is used to make the number of the minor class equivalent to the number of the major class.

5.3.4 Modelling Process

Eleven classifiers were initially used for modelling. These are Logistic Regression, Decision Tree, K-Nearest Neighbours, Random Forest, Gradient Boosting, Extreme Gradient Boost, Adaptive Boosting, Support Vector Machine, Naive Bayes, Voting Classifier and Artificial Neural Network. The classifiers giving more than 90% accuracy and higher recall are selected later for modelling purposes. Finally, six models from six classifiers are used in the modeling process. These models are then tuned

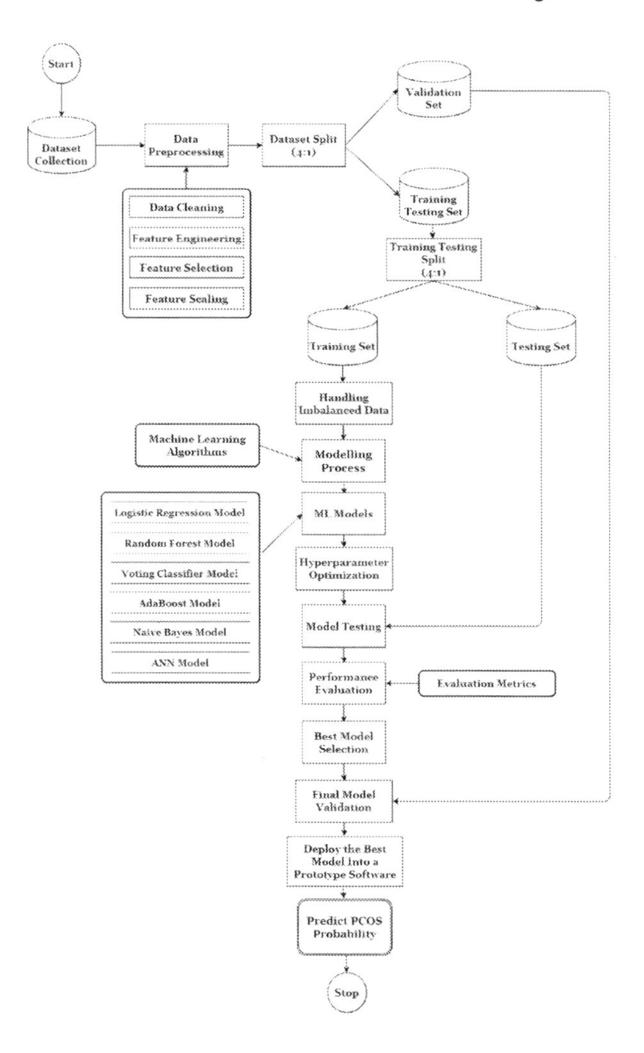

Figure 5.2 System Design

with the Grid Search hyperparameter optimization technique. The system design is given in Figure 5.2.

5.3.4.1 Hyperparameter Optimization

This is the process to maximize the model's performance by selecting a set of parameters which would give a better result because of their correct combination. This proposed system uses Grid Search, which is also known as parameter sweep. It does an exhaustive search on the specified subset of parameters, which are manually set. Cross validation is needed for guiding the method as a performance metric. As it does search for the best combination of hyperparameters from a grid of hyperparameter values, it is known as Grid Search. Cross validation is set as 10 for 10-fold cross validation in Grid Search.

5.3.4.2 Logistic Regression Classifier

This is a supervised learning classification algorithm used to predict the probability of a target variable. The presence of the target or dependent variable in the case of a binary logistic regression classification means that there are only two possible classes. With data coded as either 1 (success/yes) or 0 (failure/no), in simple words, the dependent variable is binary in nature in this case. Scikit-Learn's LogisticRegression with configuration of C = 0.1, penalty = none, solver = lbfgs, max_iter = 100 is selected as the best logistic ML model for providing higher metric scores.

5.3.4.3 Random Forest Classifier

This is an ensembled learning method where many decision trees are used for making a model. This method uses several learning algorithms to obtain better predictive validation and also merge more than one similar or different kind of algorithm for classifying data [13]. By calculating the average from all the predictions of every decision tree, the prediction is done by this classifier. Scikit-learn's RandomForest-Classifier with configuration of criterion = entropy, max_depth = 5, max_features = log2, min_samples_leaf = 5, min_samples_split = 5, n_estimators = 9 gave the best results after hyperparameter tuning.

5.3.4.4 AdaBoost Classifier

Known as an Adaptive Boosting Classifier, this is a boosting algorithm used for increasing the efficiency of classifiers which are binary. AdaBoost is a specific method of training a boosted classifier [14]. It trains a decision tree model where an equal weight is assigned every time an observation is made. The weights are increased for those observations which are hard to classify and are decreased for those which are easy to classify. The next trees are built upon the weighted data. It can combine multiple classifiers which perform poorly and can build a strong classifier based upon them. Scikit-learn's AdaBoostClassifier with configuration of n_estimators = 300, learning_rate = 0.1, algorithm = SAMME.R, base_estimator = none is selected as the best AdaBoost Model, where the base estimator is selected as the decision tree classifier by default.

5.3.4.5 Naive Bayes Classifier

As the name suggests, the algorithm utilizes the Bayes theorem (total probability theorem), which assumes that none of the features are correlated to the target class or any other features [15]. Gaussian Naive Bayes is used in simple classification problems where it makes an assumption that all the features have a normal distribution. Naive Bayes considers any particular feature in the class variable independently. The assumption Naive Bayes has is known as class conditional independence.

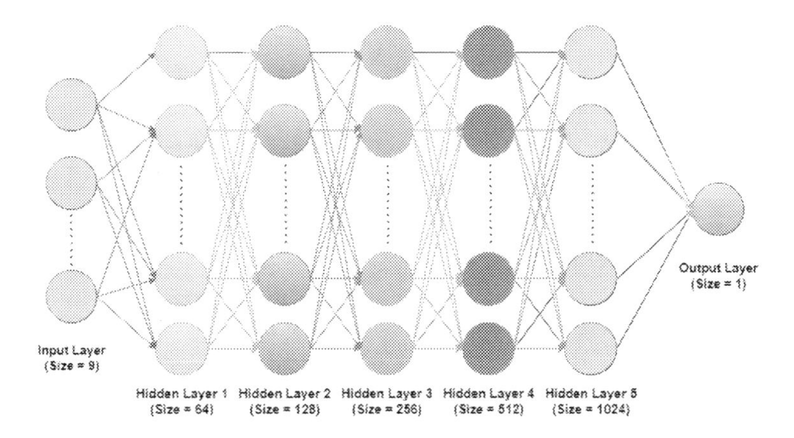

Figure 5.3 Proposed Architecture of the ANN Model

5.3.4.6 Artificial Neural Network

This algorithm is a part of deep learning. Different layers are used on the information to do mathematical processing by ANN. It tries to simulate the human brain, which is why it works on some hidden states and these hidden states' nodes are known as neurons. The proposed ANN model has five hidden layers consisting of 64, 128, 256, 512 and 1024 respective neurons for the layers with the parameters set as 100 epochs, batch size of 512, optimization algorithm set as Adam optimizer, L2 regularization technique with 0.01 weight in the loss function, and the Dropout value set as 0.4 after hyperparameter tuning. The model is shown in Figure 5.3.

5.3.4.7 Voting Classifier

This is a type of ensemble model which works like an electoral system. This algorithm can use multiple different classifiers to make a new prediction based on them. The purpose of this classifier is to create a single model which would be trained by these multiple classifiers and then the output would be predicted for all output classes based on combined majority voting. Scikit-learn's VotingClassifier, with the configuration of estimators = [Logistic Regression (Best Model), Random Forest (Best Model), Adaptive Boosting (Best Model)], voting = 'soft' where soft voting is done for prediction.

5.3.5 Performance Evaluation

Recall is the measure of correctly identifying the true positive cases. As our system works closely with human health, a patient having PCOS who is identified as not having PCOS is a devastating situation that we want to avoid. Accuracy is the metric which represents the percentage of correctly identified data points. Precision is known as the ratio which represents correctly predicted positive situations against all positive predictions. The weighted average of recall and precision is known as the F1 score.

5.3.6 Selecting Best Model

Based on the evaluation metrics, the best models created by all the classifiers would be compared and the model with the best scoring in all metrics would be selected as the final model.

5.3.7 Validating Final Model

The final model is validated by the unseen validation dataset to observe the performance of the best model.

5.3.8 Deploying Final Model into PCOS Predictor

The final model is saved and imported later with the help of the scikit learn library. PyQt5 and tkinter modules were used to build the GUI of the PCOS Predictor prototype. It takes inputs from the user and predicts the probability of a patient having PCOS.

5.4 EXPERIMENTAL RESULTS

This section includes Statistical Results and Model Visualization.

5.4.1 Statistical Results

The dataset had 44 attributes and originally had 541 patients of 20 to 48 years old. 364 patients don't have PCOS and 177 patients are PCOS affected. After feature engineering and feature selection, nine attributes are found to have more than 0.25 correlations to the class. These features have the most impact for the development of PCOS in any patient. The name of the features and how they can be obtained is shown in Table 5.1

Follicle numbers of the right and left ovaries can be found from ultrasound scans, and AMH level can be found from the AMH Hormone test. Five features are symptoms and one can be identified from lifestyle choices. These nine features will be used

Table 5.1 Selected Features and Their Description

Attributes	**Description**
Follicle numbers of right ovary	Numerical values [Scan Findings]
Follicle numbers of left ovary	Numerical values [Scan Findings]
Skin darkening	Nominal values (Yes, No) [Symptoms]
Hair growth	Nominal values (Yes, No) [Symptoms]
Weight gain	Nominal values (Yes, No) [Symptoms]
Irregular cycle of period	Nominal values (Yes, No) [Symptoms]
Fast food consumption	Nominal Values (Yes, No) [Lifestyle]
Pimple development	Nominal values (Yes, No) [Symptoms]
Anti-Mullerian Hormone level (ng/mL)	Numerical values [Hormonal Analysis]

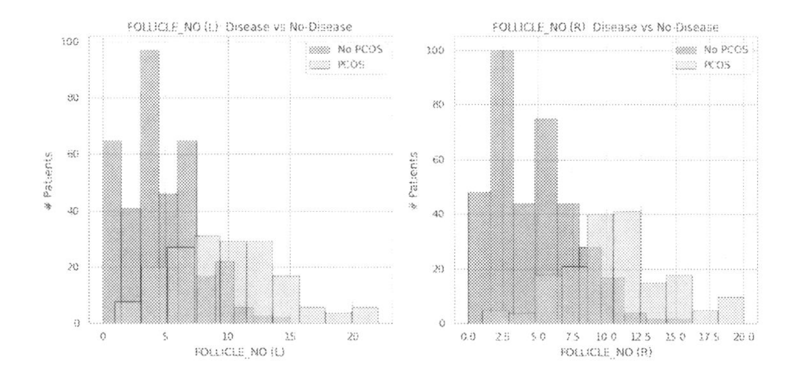

Figure 5.4 Histogram of Patients Based on Follicle Numbers

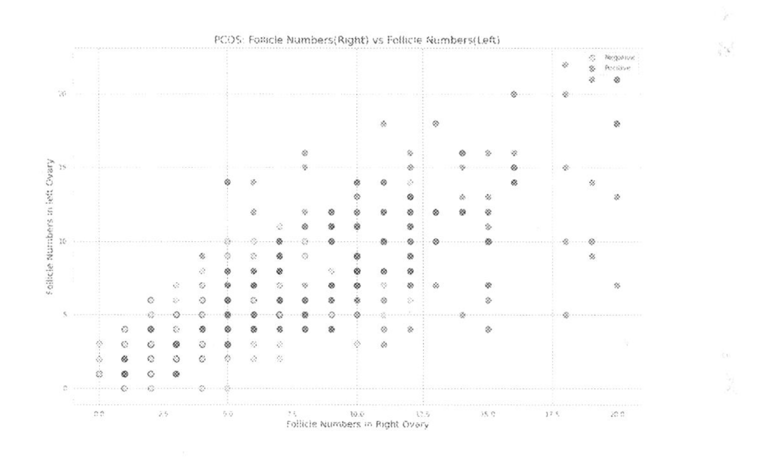

Figure 5.5 Patients' Follicle Numbers in Both Ovaries

for predicting PCOS. From further data analysis, it is found that a patient having more than 15 follicles in any ovary, will surely have PCOS. A histogram of Patients Based on Follicle Numbers is shown in Figure 5.4.

Also, if a patient has around 10 or more follicles in both ovaries, the patient has more than 90% probability of being affected by PCOS which is shown in Figure 5.5.

There are also some findings like women who are of age 20 to 35 are likely to have a high chance of having PCOS after teenage years. Follicle size doesn't matter in causing PCOS, but if the numbers of follicles are high, the patient is likely to have larger follicles, such as 14 mm and above, and be affected with PCOS. The Results Comparison is shown in Table 5.2. In this study, the Grid Search method was used to find the best combination of parameters that generated the best models for each classifier. As we observed all these evaluation metrics, the best model had to have the highest scoring combination for all these classifiers. Logistic Regression classifier had 95.4% accuracy, 94.12% recall, 94.12% precision and 95.4% F1 score. ANN, the only deep learning algorithm, also showed good results with 90.24% accuracy, 87.50% recall, 87.50% precision and 90.24% F1 score.

Table 5.2 Results Comparison

Method	Accuracy	Recall	Precision	F1 Score
Logistic Regression	95.40%	94.12%	94.12%	95.40%
Random Forest	94.25%	91.17%	93.94%	94.25%
Adaptive Boosting	91.95%	91.18%	88.57%	91.95%
Naive Bayes	90.81%	91.18%	86.11%	90.81%
Voting Classifier	93.10%	91.18%	91.18%	93.10%
Artificial Neural Network	90.24%	87.50%	87.50%	90.24%

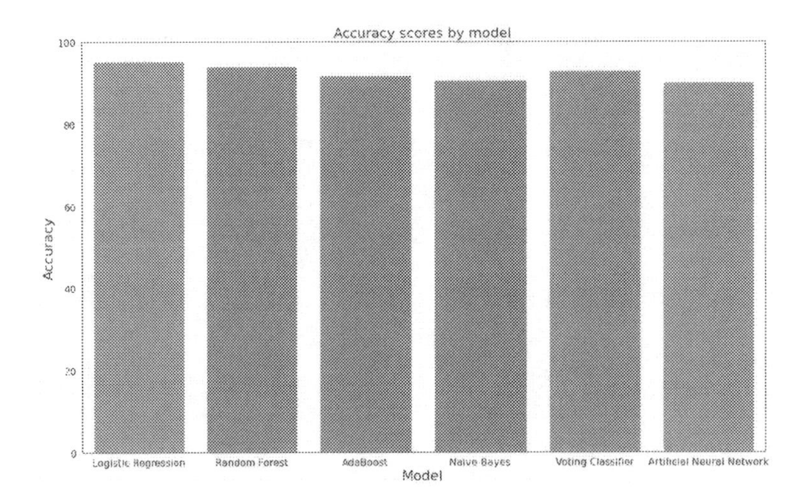

Figure 5.6 Classification Accuracy Scores of Different Models

The classification accuracy score comparison among the classifiers is shown in Figure 5.6.

As the Logistic Regression classifier has the highest recall and other scoring metrics than others, it is chosen as the final model which is hence validated with the validation set. After using the validation set, the obtained results are shown in Table 5.3.

To improve recall, a decision threshold value of 45% is used to get a higher recall value, while not sacrificing other metrics scores by a lot. This threshold value gave the best recall score while maintaining good accuracy. Different threshold values were

Table 5.3 Final Model's Validation Scoring

Metrics	Score(%)
Accuracy	94.4
Recall	87.2
Precision	97.1
F1 Score	91.9

Table 5.4 Decision Threshold Scoring

Metrics	Score(%)
Accuracy	93.5
Recall	89.7
Precision	92.1
F1 Score	90.9

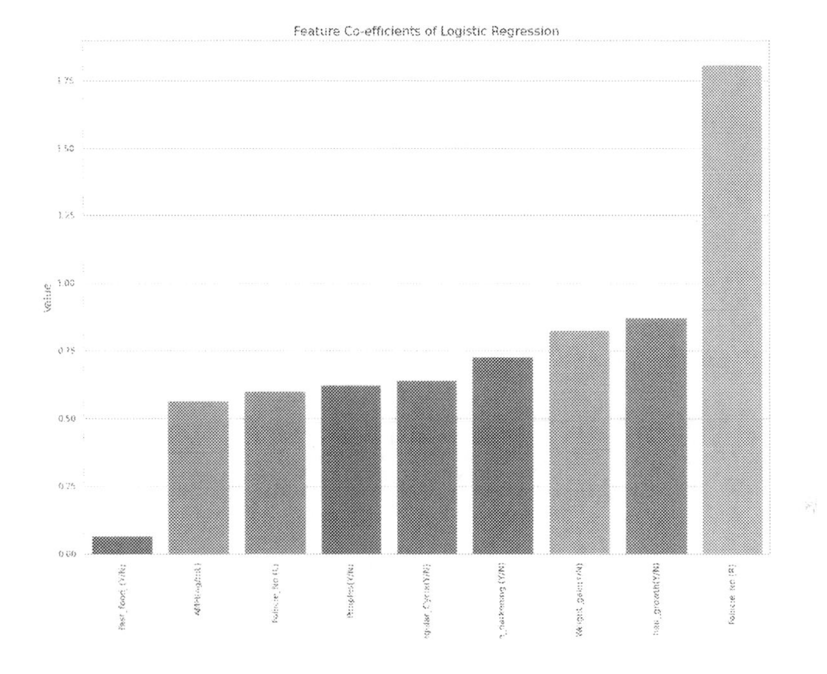

Figure 5.7 Feature Importance of Final Model

tested and this model with 45% decision threshold value can give better predictions in new data also.

This model works better than other models as this research prioritized improving the recall score to maximize the correctly predicted number of people with the disease. From this point, the research done by Amsy Denny [5] on 23 clinical and metabolic features similar to this research showed 74.19% recall with the proposed model, whereas this model gave 89.7% recall score shown in Table 5.4. Finally, this model is saved and deployed into the PCOS predictor.

The feature importance of all the nine features is also shown in Figure 5.7. Follicle numbers in the right ovary have the highest impact of 1.78, whereas hair growth has feature importance of 0.84 coming in at second place. Weight gain has 0.79, skin darkening has 0.73, irregular cycle has 0.63, pimples has 0.61, follicles in the left ovary has 0.58, AMH has 0.55 and fast-food consumption has the lowest feature importance of 0.09.

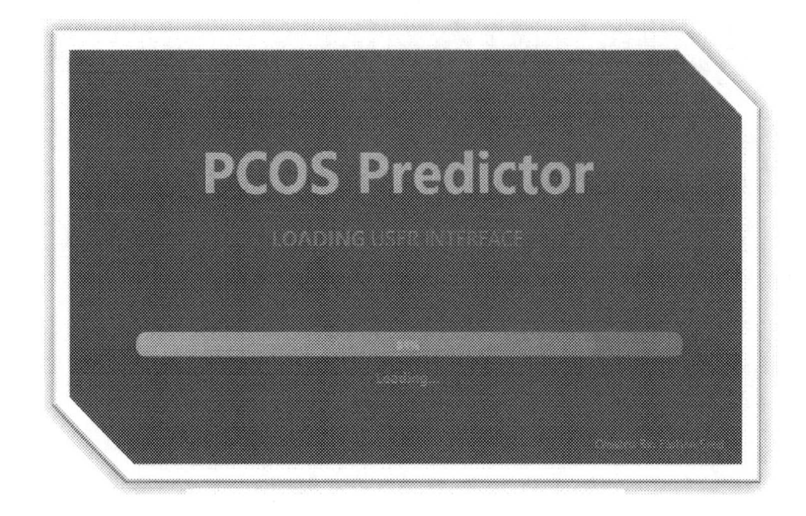

Figure 5.8 PCOS Predictor Prototype

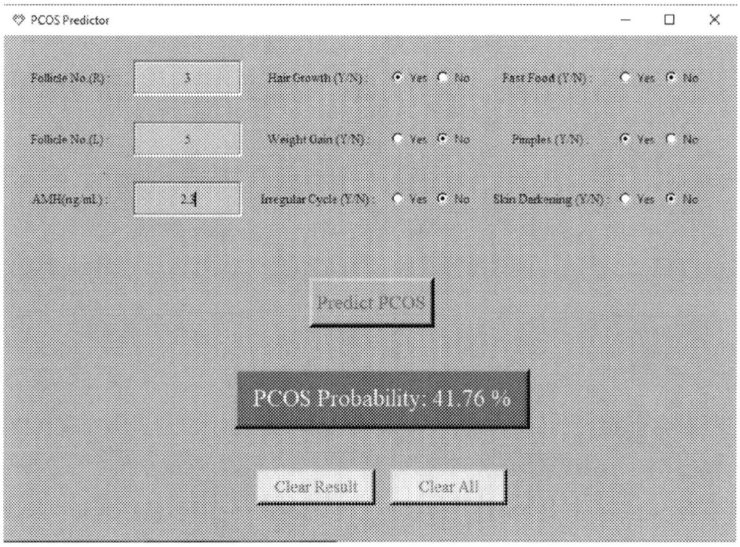

Figure 5.9 Probability Visualization (1)

5.4.2 Model Visualization

The final model is deployed into a prototype software for the purpose of visualizing new input data. Any user can give new inputs based on the nine features this research identified and the GUI will show the prediction made by the model shown as the probability of PCOS. The GUI is built with the tkinter and PyQt5 modules of Python.

This is the splash screen of the prototype which appears before loading the UI shown in Figure 5.8.

This is the user interface where the user can give new inputs and the probability is shown. PCOS probability is less than 50% here for a patient, which indicates that the patient has low probability of being affected by PCOS in Figure 5.9.

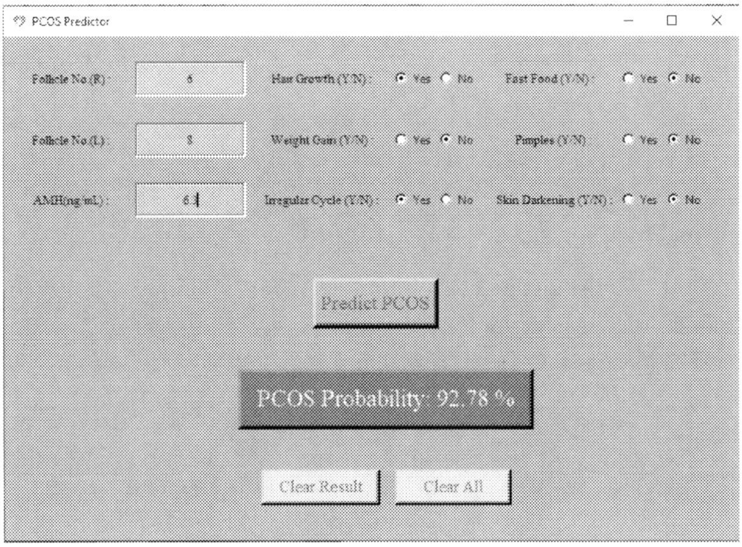

Figure 5.10 Probability Visualization (2)

In Figure 5.10, the prototype powered by the best model of logistic regression is showing that the user has a high probability of being affected by PCOS based on the most impactful nine features found for the development of PCOS.

5.5 CONCLUSION AND FUTURE WORKS

As this dataset was relatively small, we worked hard to include necessary information while removing the outliers from the equation. We achieved great accuracy and recall value while we reduced the number of features to only nine. These nine features can help doctors confirm if a patient has PCOS or not. In this study, a logistic regression classifier was chosen as the best model as it works best where the outcome is a discrete variable. The Artificial Neural Network didn't give the best metric scores because there were fewer data and it couldn't yield a better model. With the chosen model here, PCOS probability is visualized via a prototype created in Python.

We want to increase accuracy and recall of the model with collection of new data of patients as a larger dataset will surely increase the performance of the model. We want to develop the software with more functionalities. We will work with new datasets in the future to improve the model. We also wish to analyse these nine features with depth to help our sisters, and mothers who are sick and helpless with PCOS.

Bibliography

[1] Jillian, L. "Polycystic Ovarian Syndrome (PCOS) Signs and Natural Remedies." *Dr. Axe*. March 19, 2021. Accessed October 05, 2021. https://draxe.com/health/polycystic-ovarian-syndrome/.

[2] Krishnaveni, V. 2019. "A Roadmap to a Clinical Prediction Model with Computational Intelligence for PCOS." *International Journal of Management, Technology and Engineering* IX (II): 177–85.

[3] Farooque, M. "Symptoms of PCOS and Tips to Manage the Disorder Better." *Possible.* November 27, 2020. Accessed October 05, 2021. https://possible.in/pcos-symptoms-causes.html.

[4] Begum, G. S., A. Shariff, G. Ayman, B. Mohammad, and R. Housam. 2017. "Assessment of Risk Factors for Development of Polycystic Ovarian Syndrome." *International Journal of Contemporary Medical Research* 4 (1): 164–67.

[5] Denny, A., A. Raj, A. Ashok, C. M. Ram, and R. George. 2019. "I-HOPE: Detection and Prediction System for Polycystic Ovary Syndrome (PCOS) Using Machine Learning Techniques." *IEEE Region 10 Annual International Conference, Proceedings/TENCON*, 673–78. https://doi.org/10.1109/TENCON.2019.8929674.

[6] Cheng, J., and S. Mahalingaiah. 2019. "Data Mining Polycystic Ovary Morphology in Electronic Medical Record Ultrasound Reports." *Fertility Research and Practice* 7: 1–7.

[7] Dewi, R. M., Adiwijaya, U. N. Wisesty, and Jondri. 2018. "Classification of Polycystic Ovary Based on Ultrasound Images Using Competitive Neural Network Classification of Polycystic Ovary Based on Ultrasound Images Using Competitive Neural Network." In *Journal of Physics: Conference Series*, 971:1–8.

[8] Mehrotra, P., J. Chatterjee, C. Chakraborty, B. Ghoshdastidar, and S. Ghoshdastidar. 2011. "Automated Screening of Polycystic Ovary Syndrome Using Machine Learning Techniques." *Proceedings - 2011 Annual IEEE India Conference: Engineering Sustainable Solutions, INDICON-2011*, no. 1: 1–5. https://doi.org/10.1109/INDCON.2011.6139331.

[9] Zhang, X., Y. Pang, X. Wang, and Y. Li. 2018. "Computational Characterization and Identification of Human Polycystic Ovary Syndrome Genes." *Scientific Reports* 8 (1): 1–7. https://doi.org/10.1038/s41598-018-31110-4.

[10] Deshpande, S., and A. Wakankar. 2014. "Automated Detection of Polycystic Ovarian Syndrome Using Follicle Recognition." *2014 IEEE International Conference on Advanced Communication Control and Computing Technologies (ICACCCT)*, no. 978: 1341–46.

[11] Stepto, N., S. Cassar, A. Joham, S. Hutchison, C. Harrison, R. Goldstein, and H. Teede. 2013. "Women with Polycystic Ovary Syndrome Have Intrinsic Insulin Resistance on Euglycaemic – Hyperinsulaemic Clamp." *Human Reproduction* 28 (3): 777–84. https://doi.org/10.1093/humrep/des463.

[12] Cahyono, B., Adiwijaya, M. S. Mubarok, and U. N. Wisesty. 2017. "An Implementation of Convolutional Neural Network on PCO Classification Based on

Ultrasound Image." In *2017 Fifth International Conference on Information and Communication Technology (ICoICT)*, 0:3–6.

[13] Banerjee, A., A. Noor, N. Siddiqua, and M. Uddin. 2019. "Food Recommendation Using Machine Learning for Chronic Kidney Disease Patients." In *International Conference on Computer Communication and Informatics, ICCCI 2019*, 1–5. Coimbatore: IEEE. https://doi.org/10.1109/ICCCI.2019.8821871.

[14] Akter Hossain, M., and M. Uddin. 2018. "A Differentiate Analysis for Credit Card Fraud Detection." *2nd International Conference on Innovations in Science, Engineering and Technology, ICISET*, no. October: 328–33.

[15] Banerjee, A., A. Noor, N. Siddiqua, and M. Uddin. 2019. "Significance of Attribute Selection in the Classification of Chronic Renal Disease." In *2nd International Conference on Advanced Computational and Communication Paradigms(ICACCP)*, 1–6. IEEE. https://doi.org/10.1109/ICACCP.2019.8882937.

CHAPTER 6

Malware Detection: Performance Evaluation of ML Algorithms Based on Feature Selection and ANOVA

Nazma Akther

Department of Computer Science & Engineering, Premier University, Chattogram, Bangladesh

Md. Neamul Haque

Department of Computer Science & Engineering, Premier University, Chattogram, Bangladesh

Khaleque Md. Aashiq Kamal

Department of Computer Science & Engineering, King Fahd University of Petroleum and Minerals, Dhahran, Kingdom of Saudi Arabia

CONTENTS

Cyberattacks are a critical problem in the field of digital world. To mitigate cyber-attacks, malware detection in network traffic is the major concern. The success of effective malware detection approaches on network traffic mainly depends on the correctness and building time of the systems. Usually network traffic contains a huge number of packets with vast features. In this chapter, for minimizing the feature set we use four different feature selection techniques: CFS Subset eval, Consistency Subset eval, Infogain Attribute and One RA Attribute. We have found that the CFS Subset eval reduces the features of the KDD dataset to 10 from 41 features (reduction rate is above 75%). Moreover, this work compares the following machine learning approaches: Naive Bayes, J48 as Decision Tree, and Random Forest classifiers on the full and reduced featured data set. The result shows Random Forest classifier has the highest accuracy rate 99.97% with the One RA attribute selection technique. On the other hand, Naive Bayes has the lowest accuracy rate of 90.62% with the Consistency feature selection technique. Finally, we have applied the ANOVA technique to justify our experimental result statistically.

6.1 INTRODUCTION

Today, with the advancement of information and communication technology, security has become a big issue or threat to the digital world. With the use of the Internet, electronic devices like computers, mobile, tablets, ipads and laptops have become ubiquitous. The increasing number of electronic devices which are connected through the Internet raise the threat to those devices. Malicious software called malware is the most dangerous payload. It can lead to a successful cyberattack. A large number of cyberattack take place on digital platforms and include spoofing, sniffing, denial-of-service (DoS) attack or DDoS (Distributed Denial-of-Service) attacks, click fraud, User to Root attack (U2R), evil twins, pharming, probing attack, etc. The growth of cyberattacks increases dramatically and compromises the systems, taking away valuable information. On average, it costs $345 per incident by destroying valuable organization structures and producing immeasurable losses [1]. The growth of digital threats is not only due to the advancement of internet use, but also for the huge numbers of new malware. Different pieces of malware produced over 317 million in 2014 [2]. Conventional anti-virus and intrusion detection systems cannot detect zero day attacks. The circulation of malware on the internet exceeded 5 million as stated by the Symantec Internet Security Threat Report 2010 [3]. Moreover, according to the Symantec Internet Security Threat Report 2018 [4], 46% new pieces of ransomware were produced in total malware in 2017. To overcome these challenges, security specialists are working hard to create a productive malware detection method. Significant research has been conducted to develop machine learning based detection systems.

In this study, we use a benchmark network traffic dataset called the KDD dataset to identify the malware using several machine learning algorithms. We also use several filter based feature selection techniques to remove a few of features of our KDD dataset. Feature selection is an essential task for malware detection. Moreover, malware can be detected through static and dynamic features. The network traffic dataset has a lot of packets with an immense number of features. Several features may be very important for detecting malware or may not. A large number of relevant or irrelevant features may increase the processing time as well as decrease the efficiency of malware detection systems. So, analysis of different feature selection techniques is important to evaluate the result of the number of features on the precision of a system.

The following summarizes our contribution:

- Our proposed method classifies the benchmark network traffic dataset called the KDD dataset to distinguish the malware and good-ware (normal and suspicious activity) by applying three machine learning algorithms: Naive Bayes, J48 Classifier and Random Forest.
- We analyze a small feature subset that implies a trade-off between the classification accuracy and computational time to build the model.
- Finally, we use two-way ANOVA (analysis of variance) for both machine learning algorithms and feature selection techniques.

The rest of this chapter is organized as follows. Section 6.2 reviews the background details related to our work. Section 6.3 explains the problem statement of this work. Section 6.4 describes the proposed methodology of our work. Section 6.5 provides the experimental result of our analysis. Section 6.6 presents the statistical analysis of our study. Finally, Section 6.7 concludes the chapter.

6.2 RELATED WORK

There are many approaches established to represent the propagation number of malware that rise every moment. Alkasassbeh et al. [5] employed three machine learning algorithms, Support Vector Machine (SVM), BayesNet and Multi-Layer Perceptron (MLP), to detect an intruder using SNMP-MIB data from a real-time environment. For achieving better detection accuracy, they used several feature selection techniques such as GS (Generic Search) , IG (Info Gain) and RF (ReleifF), where BayesNet provides 99.9% accuracy with the GS feature selection technique.

Hansen et al. [6] offered a unique method to classify the family of malware as well as detect the unknown malware. They applied a Random Forest Classifier to identify the malware and its family with a huge number of malware and benign samples using the Cuckoo sandbox environment. Again, Tian et al. [7] proposed an automated tool running in a virtual environment to extract API call features from executables and apply pattern recognition algorithms and statistical methods to differentiate between cleanware and malware. By scrutinizing the behavioral features of the malware, the authors achieved 97% accuracy approximately. The dataset includes 456 clean-ware and 1,368 malware samples. On the other hand, to achieve the runtime behavior of malware, Darshan et al. [8] also used a Cuckoo sandbox environment. They differentiated malware by collecting data in the form of N-grams and developed a classifier with the Information Gain feature selection technique. By comparing other classifiers, they observed that SPegasos acquired a higher detection accuracy rate from various feature lengths like 600, 400 and 200.

Qiguang et al. [9] recommended an outstanding approach of bilayer abstraction embedded with sentiment analysis of API sequences for detecting malware properly. For imbalanced malware samples, they introduced a support vector machine called OC-SVM Neg, which gives a better malware detection result. The study of [10] developed an online analytical model named OADFM for analyzing, extracting, and detecting malware based on clustering of malware families. For classifying the malware families they used the LSH clustering algorithm. The rate of malware detection is almost 98%, which is extracted from shared features.

Moreover, Santos et al. [11] developed a hybrid detection method for malware detection. It combines the obtained features statically and dynamically. They used sequences of opcode,system calls, exceptions, etc., to make a feature vector. They also examined their analysis on a mixed dataset containing benign and malicious programs using SVM and Random Forest machine learning algorithms. For malware, they used VXHeaven as a source, and for the benign program, they depended on their setup. In [12], Bekerman et al. proposed a supervised system for detecting malware. They extracted 972 features of malware behavior from different observation areas. They employed Random Forest, decision tree (J48) and naive Bayes as a machine learning algorithm to make decisions. Their dataset consists of malware and benign programs. They used the dataset of Emerging Threat [13] security companies and Verint [14]. Furthermore, Ahmed et al. [15], introduced a malware detection technique combining two different dynamic features (from spatial and temporal information) available in a run-time API. They obtained 96.3% classification accuracy by using 516 executable files.

6.3 PROBLEM STATEMENT

In this work, we are trying to detect malware in a network traffic dataset like KDD by minimizing the feature using machine learning algorithms. To evaluate the accuracy of feature selection techniques, we use a classifier such as naive Bayes, J48 as Decision Tree, and Random Forest classifiers on the full and reduced feature dataset. However, to interpret the relation among feature selection techniques a machine learning algorithm will be calculated. According to our proposed work we are attempting to answer the following questions:

- Can we distinguish among machine learning approaches based on the features from the network traffic dataset?
- What type of relation do you observe among feature selection techniques and the machine learning algorithm?

Finally, we will estimate our result constructed by statistical analysis. For this research proposal, we have conveyed two types of hypothesis to be verified.

`Type 1:`

- H_a0: There is no significant difference in the variance of the feature set of the network traffic dataset among feature selection techniques.

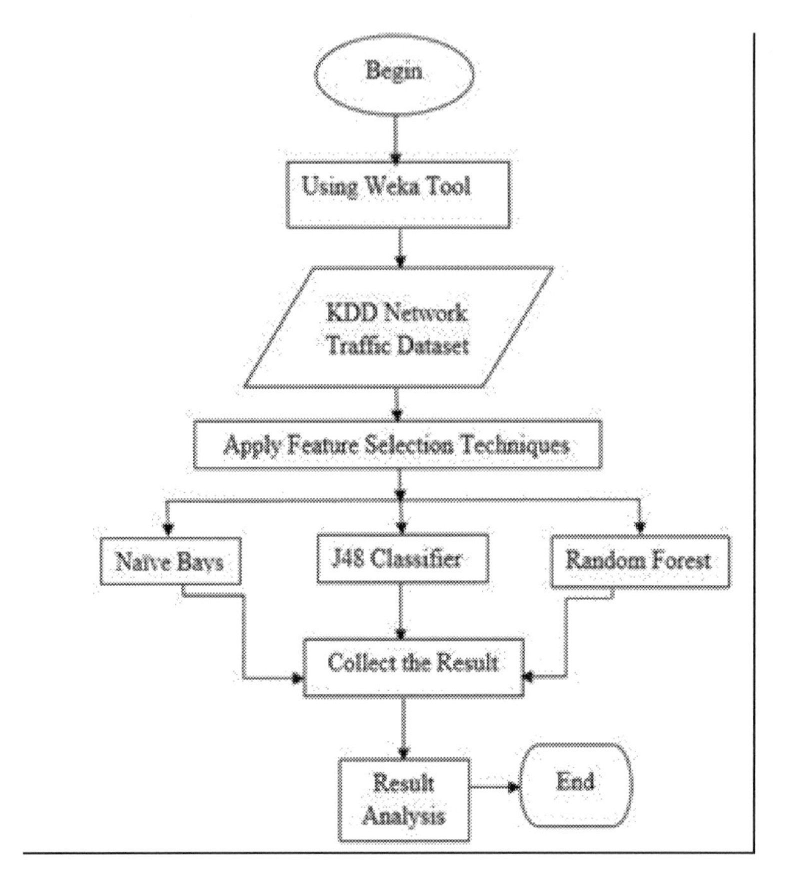

Figure 6.1 Experimental Procedure of Proposed Methodology

- H_a1: There is a significant difference in the variance of the feature set of the network traffic dataset among feature selection techniques.

Type 2:

- H_b0: There is no significant difference among the machine learning algorithms.
- H_b1: There is significant difference among the machine learning algorithms.

6.4 RESEARCH METHODOLOGY

In the process of achieving our objective, we used several machine learning algorithms for distinguishing malware or benign samples. The purpose of our proposed work is to categorize the malware by evaluating the effectiveness and accuracy of the classifier on the network traffic dataset. Our experiments were performed with real malware samples collected from the KDD dataset [16]. The experimental procedure of our proposed method is shown in Figure 6.1. Here, an open-source data mining tool, Weka, is used to carry out the study. During this study different combinations of experiments are performed with a full and reduced featured data set of KDD using three machine learning algorithms. We apply filter based feature selection techniques

Table 6.1 Environmental Setup for our experiments

Topic	Used Element
Model	HP ProBook Laptop
Operating System	Windows 7
Processor	Intel Core i5 2.4 GHz
RAM	4096 MB
Internet	Not connected

to reduce the number of features from the KDD data set. Finally, we collected all the experiment results and used analysis of variance (ANOVA) to verify the significance of feature selection techniques and machine learning algorithms. Table 6.1 shows the environmental setup of our proposed method.

6.4.1 Data set

In the field of intrusion detection techniques, the KDD dataset is a well-known benchmark dataset. To create this dataset, a large volume of network traffic such as e-mail, web, and Telnet were captured and generated. A simulation is made of categorizing [17] four different types of attack like Denial of Service (Dos), Remote to Local (r2l), User to Root (u2r) and Probe by using various operating systems. There are 42 attributes in the KDD dataset. Forty-one out of these 42 attributes can be categorized into four classes such as Basic (B) features for TCP connection, Content (C) features for domain Knowledge, Traffic (T) features for computing two-second time window and Host (H) features for assessing attacks which last for more than two seconds.

6.4.2 Weka Tool

To achieve our goal, we use Waikato Environment for Knowledge Analysis (Weka) [18], an available data mining tool under GNU. In this study we used version 3.7.11, which has many machine learning algorithms for data analysis. The Weka tool accepts data files either in attribute-relation (arff) or comma-separated value (csv) file format.

6.4.3 Feature Selection Technique

In a malware detection system, one should develop an efficient solution to decrease the number of data dimensions. Feature selection is a way to reduce the computation time by reducing model training time. It reduces the data dimension size that is needed for training the model efficiently [21]. In a simplified manner, the feature selection technique keeps the most relevant features and eliminates irrelevant and redundant ones. It reduces over-fitting by reducing redundant features in the dataset [22]. There are mainly three types of feature selection techniques: the filter method, wrapper method, and hybrid methods. The filter method is more suitable than other methods when managing a large data set [21]. As a result, we applied four different feature selection techniques based on the filter method. In the filter method, a rank score is given to each feature in the working data set. After that, they are ordered based on their score. Finally, features with higher rank are considered. To analyze the effect on accuracy of three machine learning algorithms, naive Bayes, J48 classifier and Random Forest are considered. During our work, a total fifteen individual tests are carried out with arrangement of feature selection techniques and machine learning approaches.

6.5 RESULT ANALYSIS

In this experiment, we reduced features of the dataset by using different feature selection techniques based on the filter method and used mainly three different machine learning algorithms: Naive Bayes, the J48 decision tree, and Random Forest. Fifteen different combinations of experiments have been performed in this work based on feature selection techniques and machine learning algorithms. The following feature selection techniques have been used:

- CFS Subset Eval: The number of features of the examined dataset are reduced to 10 from 41 features using this technique. It is an enormous improvement with a 75% reduction in the feature set.

- Consistency Subset Eval: By using Consistency Subset Eval, we reduced the number of features of the examined dataset to 14 from 41 features. It is also a good improvement with a 66% reduction in the feature set.

- Info Gain Attribute: We get 30 features of the examined dataset from 41 features using this technique. In this case, we achieve only 27% improvement.

- One RA Attribute: We get 30 features from the examined dataset from 41 features using this technique. In this case, we achieve only 27% improvement.

Here, Figure 6.2 shows the reduction rate of features using different feature selection techniques (in percentage). We can see that CFS Subset eval reduced the features 75% from the full feature set. In contrast, Info gain attribute and One RA attribute techniques achieved the lowest reduction rate of only 27%.
For all experiments, 10-fold cross validation is used for testing the models. The 10-fold cross validation reduces the variance of the estimation [23]. When we train the KDD dataset with complete featured means considering 41 attributes, the Naive Bayes classifier provides 92.81% accuracy with 7.30 sec, J48 provides 99.96% accuracy with 96.68 sec, and the Random Forest classifier provides 99.98% accuracy with computational time of 898.47 sec, along with a full features dataset. On the other hand, if we filter the feature or shrink the number of features with feature selection techniques, the time computation will be reduced. Table 6.2 shows the CFS subset eval has the lowest with only 0.79 sec of computation time. The rate of accuracy is 92.71%, which is very close to the full featured accuracy rate. As the CFS feature selection technique works with ten features, it reduces almost 75% of the features. Table 6.3 for the J48 classifier also shows that when we filter the features applying the feature selection technique, the time computation will be reduced, the algorithm

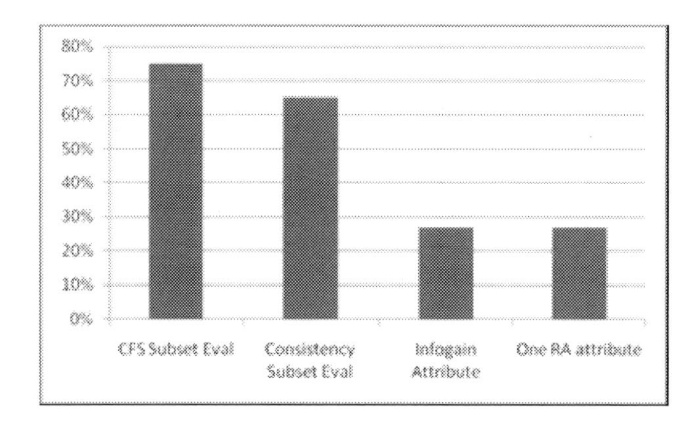

Figure 6.2 Reduction rate of features using different feature selection techniques (in percentage)

Table 6.2 Accuracy on different feature selection techniques using Naive Bayes Classifier

Feature selection Technique	Build Model	Accuracy	F-measure
CFS Subset Eval	0.79 Sec.	92.71%	0.934
Consistency Subset Eval	1.05 Sec	90.62%	0.933
Infogain attribute	2.18 Sec	91.41%	0.939
One RA attribute	2.44 Sec	92.73%	0.950
Full Dataset	7.30 Sec	92.81%	0.955
Average	**2.75 Sec**	**92.06%**	**0.942**

CFS subset eval takes 14.12 sec, while the consistency subset takes 22.02 sec to build the model with an accuracy rate of 99.98%. The accuracy rate of CFS subset eval is near enough to the full features accuracy rate. Table 6.4 for the Random Forest classifier indicates that it reduced the computational time to build the model with higher accuracy using a minimum number of features. Table 6.2, Table 6.3 and Table 6.4 show that the CFS feature selection techniques take 36.3 sec where accuracy is 99.90%. Figure 6.3 and Figure 6.4 depict average accuracy rate (in percentage) and building time of the models (in seconds) of three examined machine learning approaches. From Figure 6.3 we can say that the accuracy of J48 and Random Forest are noticeably better than the naive Bayes algorithm. J48 and Random Forest are both based on a decision tree algorithm. J48 is a simple implementation of the C4.5 decision tree algorithm and Random Forest is an ensemble technique which is a set of combined decision trees. In general, decision tree based classifiers reduce the biasness in the data set and it works better with the larger data set. As a result, we achieved higher accuracy in percentage and F-measure with the J48 and Random Forest classifier. However, from naive Bayes is more efficient than J48 and Random Forest. It is clearly shown in Figure 6.4 that the average building time of naive Bayes model is significantly lower than other classifiers.

Table 6.3 Accuracy on different feature selection techniques using J48 Classifier

Feature selection Technique	Build Model	Accuracy	F-measure
CFS Subset Eval	14.12 Sec.	99.88%	0.99
Consistency Subset Eval	22.02 Sec	99.93%	0.99
Infogain attribute	51.22 Sec	99.95%	1.00
One RA attribute	58.38 Sec	99.94%	0.99
Full Dataset	96.68 Sec	99.96%	1.00
Average	**48.48 Sec**	**99.93%**	**0.99**

Table 6.4 Accuracy on different feature selection techniques using Random Forest Classifier

Feature selection Technique	Build Model	Accuracy	F-measure
CFS Subset Eval	36.30 Sec.	99.90%	0.99
Consistency Subset Eval	48.21 Sec	99.92%	0.99
Infogain attribute	410.13 Sec	99.96%	1.00
One RA attribute	431.31 Sec	99.97%	1.00
Full Dataset	898.47 Sec	99.98%	1.00
Average	**364.88 Sec**	**99.95%**	**0.996**

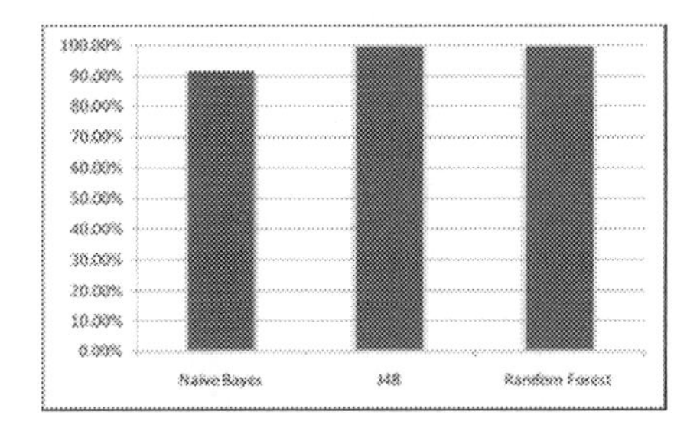

Figure 6.3 Average accuracy rate of three machine learning techniques (in percentage)

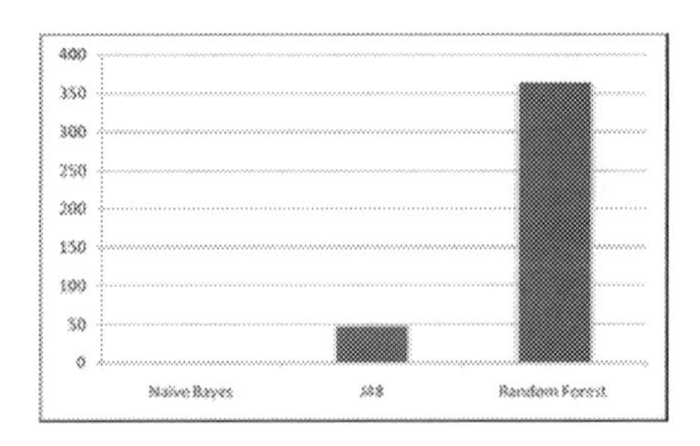

Figure 6.4 Average building time of three machine learning models (in seconds)

Table 6.5 ANOVA Analysis Table

Variation Source	S_mS_q	d_gf	M_nS_q	F
Between Technique(B)	$\sum n_j(u_j - V)^2$	$d_gf_1 = k - 1$	$\frac{S_mS_qB}{d_gf_1}$	$\frac{M_nS_qB}{M_nS_qE}$
Error (E)	$\sum\sum(X - u_j)^2$	$d_gf_2 = N_t - k$	$\frac{S_mS_qE}{d_gf_2}$	–

6.6 STATISTICAL ANALYSIS

In this section we have included statistical analysis to analyze our obtained experimental results. From the problem statement section, it is clear that two types of hypotheses need to be verified. For verification purposes, we have done ANOVA (Analysis of Variance) technique [19]. It is a statistical technique that is applied to distinguish different factors to perform analysis of variance [20]. The major aim of ANOVA is to statistically validate the alternatives. In this work, we have applied two-way ANOVA without replication to analyze the experimental results of Tables 6.2, 6.3, and 6.4 to verify H_a and H_b. The ANOVA technique uses the F-test, which is decompressed to F-calculated and F-tabulated. Table 6.5 shows how F-calculated is obtained. In Table 6.5 we consider d_gf as a degree of freedom, S_mS_q is expressed as sum of square, M_nS_q means mean square and F represents the test static. Moreover, V, U_j, N_t, k and X represent Overall sample mean, Sample mean of the jth group, Total sample size, independent comparison groups and individual observation, respectively. From Tables 6.2, 6.3, and 6.4 we can understand that the value of $k = 5$ and $N_t = 15$.

6.6.1 Statistical Analysis of Feature Selection Technique

According to the ANOVA technique we are setting up two hypotheses, the null hypothesis H_0, and the alternative hypothesis H_1, and the level of significance is $Alpha = 0.05$.

- H_0: Means are all equal
- H_1: Means are not all equal

Now $d_gf_1 = k - 1 = 5 - 1 = 4$ and $d_gf_2 = N_t - k = 15 - 5 = 10$. By using these two values we get the tabulated value of F (F-tabulated = 3.478). After that, we compute the calculated value of F for the feature selection techniques using the formulas represented in Table 6.5. Finally, from the resulting Table 6.6, it is clear that F-Calculated < F-Tabulated, i.e., 0.016 < 3.478. As the calculated F value of Feature selection Techniques (0.016) is less than its tabulated value, the null hypothesis H_0 is not rejected. Therefore, it is statistically proved that *"There is no significant difference among feature selection techniques."*

Table 6.6 Calculated value of F for feature selection techniques

Variation Source	S_mS_q	d_gf	M_nS_q	$F - Calculated$
Between Technique(B)	$S_mS_qB = 1.33$	$d_gf_1 = 4$	$M_nS_qB = 0.33$	0.016
Error (E)	$S_mS_qE = 209.75$	$d_gf_2 = 10$	$M_nS_qE = 20.97$	

Table 6.7 Calculated value of F for machine learning algorithms

Variation Source	S_mS_q	d_gf	M_nS_q	$F - Calculated$
Between Technique(B)	$S_mS_qB = 207.14$	$d_gf_1 = 2$	$M_nS_qB = 103.57$	315.569
Error (E)	$S_mS_qE = 3.94$	$d_gf_2 = 12$	$M_nS_qE = 0.33$	

6.6.2 Statistical Analysis of Machine Learning Algorithm

Similarly, we compute the calculated value of F for the machine learning algorithms. In this case, $d_gf_1 = k - 1 = 3 - 1 = 2$ and $d_gf_2 = N_t\text{-}k = 15 - 3 = 12$. Then we get the tabulated value of F (F-tabulated= 4.103). At last, we compute the calculated value of F using the formulas represented in Table 6.5. The calculated value of F is represented in Table 6.7. As 315.569 > 4.103, i.e., F-calculated > F-tabulated (critical value), so we can reject H_0 (null hypothesis). Therefore, after statistical analysis we arrive at the decision, ,*"There is significant difference among the machine learning algorithms."*

6.7 CONCLUSION

This work evaluated four different filter based feature selection techniques to reduce the number of features in the KDD dataset. According to the outcome, we can say that the CFS subset eval technique chooses the minimum number of features, whereas the Info gain attribute technique chooses the maximum number of features. Then we examine three different machine learning approaches using the full and reduced featured dataset. From the experiments, we can conclude that Random Forest has the highest efficiency with the accuracy rate of 99.95% on an average. But considering the building time of the model, the Naive Bayes algorithm is most efficient with the lowest model building time, which is only 2.75 seconds on average. According to the ANOVA statistical analysis, the variance of the features set of the data set among the feature selection techniques are not significant. Though it shows that there is a significant difference among the machine learning algorithms statistically, the experiment results presented in the chapter clearly show that the average accuracy of the J48 and Random Forest classifiers is almost the same, whereas the naive Bayes classifier performed significantly worse, which mainly contributed in the variance.

Bibliography

[1] Hu, X. (2011).*Large-scale malware analysis, detection, and signature generation* (Doctoral dissertation, University of Michigan).

[2] Cepeda, C., Tien, D. L. C., & Ordónez, P. (2016, October). Feature selection and improving classification performance for malware detection. *In 2016 IEEE International Conferences on Big Data and Cloud Computing (BDCloud), Social Computing and Networking (SocialCom), Sustainable Computing and Communications (SustainCom)(BDCloud-SocialCom-SustainCom)* (pp. 560–566). IEEE.

[3] Fossi, M., & Blackbird, J. (2011). Symantec internet security threat report 2010. *Symantec Corp.*, Mountain View, CA, USA, Tech. Rep.

[4] Cleary, G., Corpin, M., Cox, O., Lau, H., Nahorney, B., O'Brien, D., ... & Wueest, C. (2018). Internet security threat report. *Symantec Corporation*, California, USA, Tech. Rep.

[5] Alkasassbeh, M. (2017). An empirical evaluation for the intrusion detection features based on machine learning and feature selection methods . arXiv preprint arXiv:1712.09623.

[6] Hansen, S. S., Larsen, T. M. T., Stevanovic, M., & Pedersen, J. M. (2016, February). An approach for detection and family classification of malware based on behavioral analysis. *In 2016 International Conference on Computing, Networking and Communications (ICNC)* (pp. 1–5). IEEE.

[7] Tian, R., Islam, R., Batten, L., & Versteeg, S. (2010, October). Differentiating malware from cleanware using behavioural analysis. *In 2010 5th International Conference on Malicious and Unwanted Software* (pp. 23–30). IEEE.

[8] Darshan, S. S., Kumara, M. A., & Jaidhar, C. D. (2016, December). Windows malware detection based on cuckoo sandbox generated report using machine learning algorithm. *In 2016 11th International Conference on Industrial and Information Systems (ICIIS)* (pp. 534–539). IEEE.

[9] Miao, Q., Liu, J., Cao, Y., & Song, J. (2016). Malware detection using bilayer behavior abstraction and improved one-class support vector machines. *International Journal of Information Security*, 15(4), 361–379.

[10] Wang, Y., Chen, S., Liang, B., Song, Y., Xiao, X., & Kang, X. (2016, October). Online analytical and detection model of familial malware based on communication features. *In 2016 9th International Congress on Image and Signal Processing, BioMedical Engineering and Informatics (CISP-BMEI)* (pp. 1990–1994). IEEE.

[11] Santos, I., Devesa, J., Brezo, F., Nieves, J.,& Bringas, P. G. (2013). Opem: A static-dynamic approach for machine-learning-based malware detection. *In International Joint Conference CISIS'12-ICEUTE´ 12-SOCO´ 12 Special Sessions* (pp. 271–280). Springer, Berlin, Heidelberg.

[12] Bekerman, D., Shapira, B., Rokach, L., & Bar, A. (2015, September). Unknown malware detection using network traffic classification. *In 2015 IEEE Conference on Communications and Network Security (CNS)* (pp. 134–142). IEEE.

[13] "Emerging Threats": http://www.emergingthreats.net.

[14] "Verint," [Online]. Available: http://www.verint.com.

[15] Ahmed, F., Hameed, H., Shafiq, M. Z., & Farooq, M. (2009, November). Using spatio-temporal information in API calls with machine learning algorithms for malware detection. *In Proceedings of the 2nd ACM Workshop on Security and Artificial Intelligence* (pp. 55–62).

[16] KDDdataset, https://kdd.ics.uci.edu/databases/kddcup99 [Accessed 20 August, 2018].

[17] Aggarwal, P., & Sharma, S. K. (2015). Analysis of KDD dataset attributes-class wise for intrusion detection. *Procedia Computer Science*, 57, 842–851.

[18] Weka: Data mining software, https://www.cs.waikato.ac.nz/ml/weka [Accessed 20 August, 2018]

[19] Jain, R. (1990). The Art of Computer Systems Performance Analysis: Techniques for experimental design, measurement, simulation, and modeling. *John Wiley & Sons.*

[20] Kamal, K. M. A., Alfadel, M., & Munia, M. S. (2016, December). Memory forensics tools: Comparing processing time and left artifacts on volatile memory. *In 2016 International Workshop on Computational Intelligence (IWCI)* (pp. 84–90). IEEE.

[21] Alkasassbeh, M. (2017). An empirical evaluation for the intrusion detection features based on machine learning and feature selection methods. arXiv preprint arXiv:1712.09623.

[22] Guyon, I., & Elisseeff, A. (2003). An introduction to variable and feature selection. *Journal of Machine Learning Research*, 3(Mar), 1157–1182.

[23] Ghojogh, B., & Crowley, M. (2019). The theory behind overfitting, cross validation, regularization, bagging, and boosting: Tutorial. arXiv preprint arXiv:1905.12787.

CHAPTER 7

An Efficient Approach to Assess the Soil Quality of Sundarbans Utilizing Hierarchical Clustering

Diti Roy

Institute of Information and Communication Technology
Khulna University of Engineering & Technology (KUET)

Md. Ashiq Mahmood

Institute of Information and Communication Technology
Khulna University of Engineering & Technology (KUET)

Tamal Joyti Roy

Institute of Information and Communication Technology
Khulna University of Engineering & Technology (KUET)

CONTENTS

SUNDARBANS located in South Asia on the border between India and Bangladesh is the biggest mangrove forest in the world which is distinctive in its characteristics from others. In this study, an approach has been made to assess the soil quality of Sundarbans. To conduct our study, we have collected samples from five different regions of Sundarbans, Bangladesh, such as Sutarkhali, Karomjol, Harbaria, Akram Point and Alorkol. As soil layers are divided into four types of layers according to soil depth, we have collected samples from every layer from each location. In our study, we have measured different soil nutrients such as Organic Carbon, Nitrogen, Bulk Density, pH, Electrical Conductivity and Phosphorous. To assess soil nutrients and to exhibit correlation between them, we have used Pearson correlation coefficient along with other different statistical tools and finally, we have used a hierarchical clustering algorithm to find out the region that has the same characteristics according to these soil nutrients. Our study shows the presence of Organic Carbon is 1.84% to 2.25%, Bulk Density 1.12 to 1.21 g/cm^3, Nitrogen 0.59 to 0.69 mg in per gm, phosphorous 0.348 to 0.389 mg in per gm, pH value is 7.13 to 7.19, electrical conductivity is 2.15 to 5.63 ms in per cm in different layers of soil and last of all, our study shows the area having the same properties according to these soil nutrients.

7.1 INTRODUCTION

Forests are important globally because of their capacity to provide goods and services to people through assistance in diminishing carbon dioxide, reducing soil contamination, ensuring safety to watersheds and maintaining mountain areas [1]. Among the different types of forest, mangrove forests are unique in nature for different characteristics. Mangroves are tidal forests that are situated in sheltered coastlines, estuaries in both tropical and subtropical regions, where they are regulated uniquely by tides, salinity, rainfall, etc. [2]. Mangrove forests are essential for their assistance in growing different types of terraneous and marine flora and fauna, fish, along with several animal species including mammals, amphibians, etc. [3]. Although mangrove forests currently exist, day by day they are dwindling by 1% to 2% percentage in every year because of irresponsible decisions by people [4]. Sundarbans, which lies in the southwestern part of Bangladesh, is the largest mangrove forest in the world. About two centuries ago it was estimated that the area of Sundarbans was 16000 square kilometers, but it has been reduced to approximately half of its original size. Now, the area covered by Sundarbans is 10000 square kilometers and two-thirds of it is located in Bangladesh [5]. A magnificent variety is observed in mangrove forest land compared to other land in terms of the difference in soil nutrients such as pH, bulk density, carbon, salinity and different types of organic and inorganic particles [7]. Among these nutrients, excessive presence of organic carbon and salinity is affecting the soil quality of Sundarbans more than other nutrients [6]. The purpose of cluster analysis is to find a system of observations that implies that entities with homogeneous characteristics are members of the same group, or cluster, but when compared to other groups, their characteristics are heterogeneous [8]. Though Sundarbans is the largest mangrove forest in the world there has been limited study of the soil characteristics of Sundarbans. The objective of our study is to assess the soil quality of every Sundarbans location that we used for our study according to different soil nutrients in different layers of soil and finally, we have tried to identify the region which carries similar properties depending by these soil nutrients.

7.2 RELATED WORK

Using clustering algorithms in assessment of soil pollution is an important tool. Soil quality is influenced by human activities in the form of industrialization, agricultural activity, etc. [9] examined the soil quality of a mangrove forest by using a support vector machine. They found that anthrophonic activity is making a significant change in soil quality by altering the geochemical and hydrological characteristics of the mangrove forest. They also said that climate change is severely affecting mangrove forests of the world. Mangrove forests contain the highest amount of carbon in the existing ecosystem. [10] conducted a study with a view to calculating the loss of carbon from the land surface. By analyzing data from 1996 to 2016 they found that soil carbon stock has been reduced by deforestation of mangrove forests. At least 1.8 percent carbon lessened in last 20 years. [11] also found same result. They concluded that at least 30-122 Tg of carbon has been lost from upper part of soil due in last 20 years. At least 70% of carbon has been lost from Indonesia, Malaysia and Myanmar, whereas Bangladesh had lost lowest carbon from mangrove forests. Using hierarchical cluster analysis, a study was conducted by [12] to assess the water quality of the Varuna River in India. They also used Pearson correlation to find the magnitude of the relationship between the parameters. By analyzing samples from 10 rivers they concluded that DO, Nitrate, BOD, COD and Alkalinity were above the government recognized unit. From correlation analysis they found that pH, electrical conductivity, total alkalinity and nitrate have an impact on concentrating other water parameters. Cluster analysis helped them to identify the best result. In order to increase agricultural productivity analysis of soil quality is essential. [13] conducted a study to delineate the soil management zone by using a soil quality index. They made the study to test the impact of rice cultivation on soil nutrients of mangrove forests. That's why they collected samples from 48 points including both mangrove forest and agricultural fields close to the mangrove forest to develop a soil quality index by measuring several soil nutrients such as organic carbon, potassium, soil pH, ammonium oxidizer, phosphate and micro nutrients (Zn, Mg) etc. They found that the value of the soil index is higher in the mangrove forest than adjacent agricultural fields in every sample source. A study was conducted in the East Nile Delta zone by [14] to exhibit site specific management zones by fuzzy clustering algorithms. In their research process, they collected electrical conductivity through an EM38 sensor from 432 locations and other soil attribute such as pH, soil saturation rate, organic elements such as nitrogen, phosphorous, and calcium carbonate from 80 locations. An ordinary Kriging geo-statistical method was used to find the maps of spatial variability of soil, whereas principal component analysis was used to find the most significant attribute of soil for representing within-field variability. Finally, they concluded that soil saturation and the digital elevation model is important soil attributes and from analysis of fuzzy clustering algorithms, they found no need to classify the soil for more than five regions.

7.3 PROPOSED METHODOLOGY

Study Area

Our study area is Sundarbans, Bangladesh which is located in the latitudes 21 degrees 30 minutes N and 22 degrees 30 minutes N, and longitudes 89 degrees 00 minutes E and 89 degrees 55 minutes E.

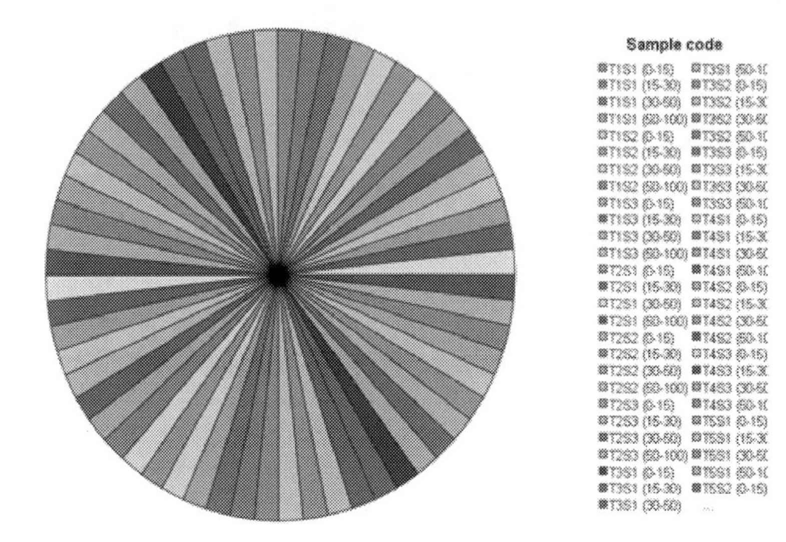

Figure 7.1 Sample Code of Different Regions

Table 7.1 Sample Point

Sample Code	Name
T1S1-T1S3	Sutarkhali
T2S1-T2S3	Karomjol
T3S1-T3S3	Harbaria
T4S1-T4S3	Akram Point
T5S1-T5S3	Alorkol

Data Sampling and Measurement

To conduct our study, we collected soil samples from 5 points of Sundarbans in Bangladesh with the help of the Bangladesh forest department. Every point is classified as 4 parts according to soil depth which is shown in Figure 7.1. Table 7.1 describes the sample code and different regions of our study within Sundarbans.

In Table 7.1, T_n indicates the area name of our collected data, whereas S_n means a sub-point within the same region. As soil layers are classified into four steps, we collected data from every layer to assess the soil quality of the region.

First, the samples of soil have been collected from different places in the Sundarbans with the help of the Bangladesh forest department like Sutarkhali, Karomjol, Harbaria, Akram point and Alorkol. All samples have been collected for every layer like 0-15,15-30,30-50 and 50-100. Then the samples have been tested to find the different values for different nutrients of soil like Bulk Density, Organic Carbon, pH, EC, N and P. After finding these values, all data have been preprocessed for analysis. Pearson correlation has been applied to find out the relation between nutrients, and different statistical tools have been used to find the mean value of Bulk Density, Organic Carbon, pH, EC, N and P of every layer of soil depth to assess the soil quality of this region of the Sundarbans. Finally, Hierarchical Clustering was used to

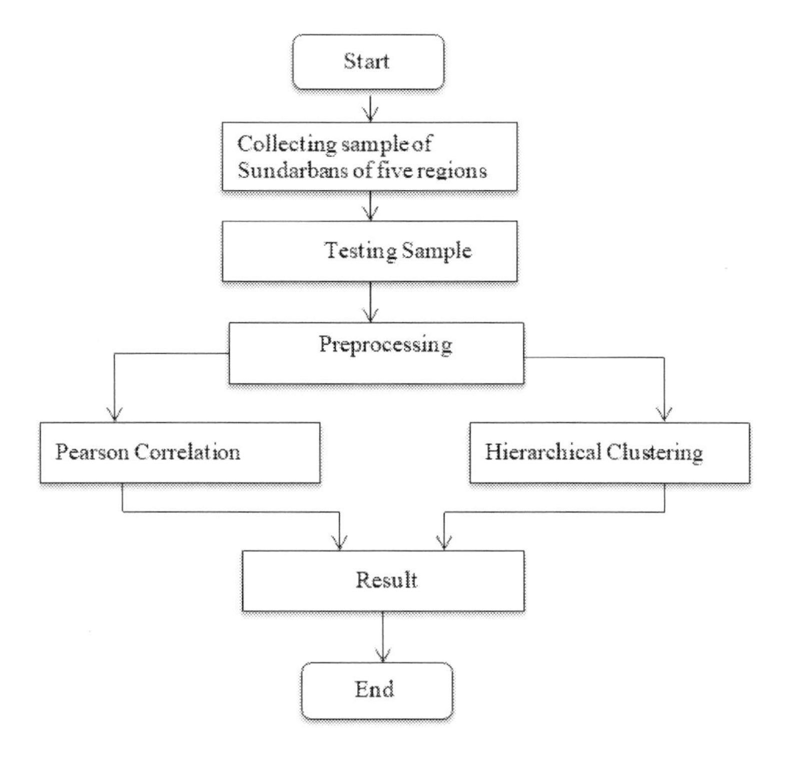

Figure 7.2 Overall working procedure

show which region has the same characteristics as the Sundarbans soil nutrients. The overall working procedure is shown in Figure 7.2.

Correlation: Correlation is an important statistical tool to show the extent of the relation between two variables. The coefficient of correlation shows the strength of the relation with each other. A positive correlation means that if one variable increases, the other variable will increase, whereas a negative correlation means that an increase in an independent variable decreases the other variable. There are many types of correlation. In our study we have used the Pearson correlation coefficient. The Pearson coefficient correlation formula is:

$$r = \frac{\sum(x_i - \bar{x})(y_i - \bar{y})}{\sqrt{\sum(x_i - \bar{x})^2 \sum(y_i - \bar{y})^2}} \tag{7.1}$$

r =correlation coefficient
x_i =values of the x-variable in a sample
$\bar{x}$ =mean of the values of the x-variable
y_i =values of the y-variable in a sample
$\bar{y}$ =mean of the values of the y-variable

Clustering: Clustering is an unsupervised machine learning approach which is used to analyze the grouping data. From a given data set, we can classify the data in a specific group by using clustering algorithms. We can interpret that data in the same cluster hold the same characteristics and they are different from other groups. Among different machine learning algorithms we have used hierarchical clustering in

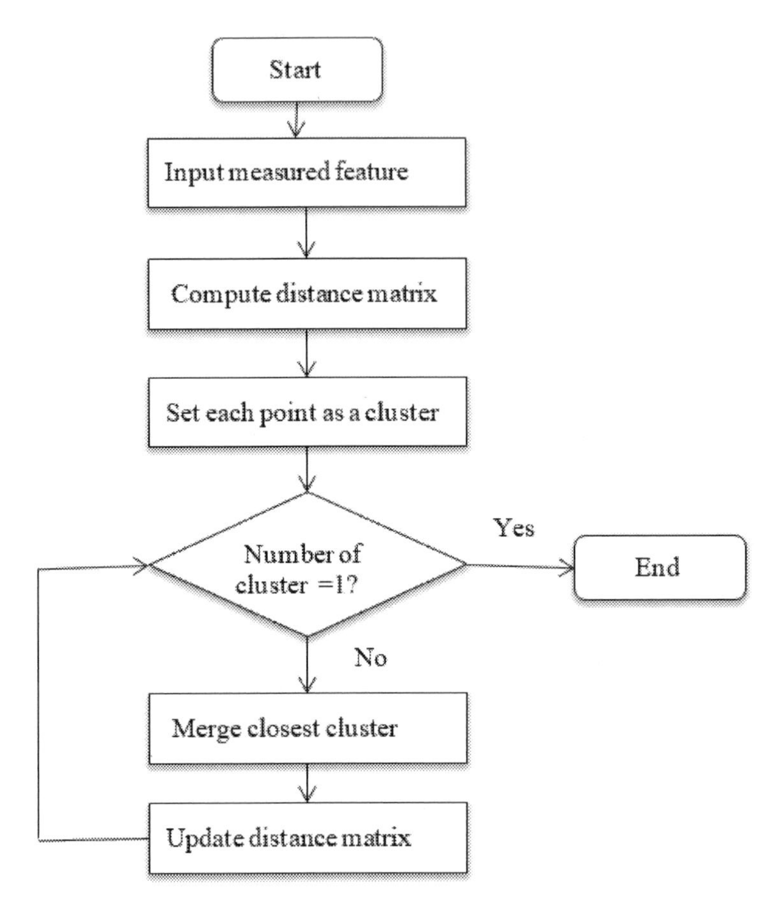

Figure 7.3 Flowchart of Hierarchical Clustering

Table 7.2 Mean Value of Soil Nutrients.

Soil Depth	Organic Carbon	Bulk Density	N	P	EC	pH
0-15	1.84	1.12	0.59	0.348	2.15	7.13
15-30	2.25	1.21	0.69	0.389	3.86	7.16
30-50	2.01	1.23	0.64	0.381	4.32	7.33
50-100	2.09	1.19	0.60	0.375	5.16	7.13

our study. Hierarchical clustering is used to classify similar data into similar groups where every cluster is different from others, but the data within each cluster is similar. The working procedure of Hierarchical Clustering of our study is shown in Figure 7.3.

7.4 RESULTS AND DISCUSSION

For this analysis, we have completed different steps. First of all, we have collected data from different regions of Sundarbans and preprocessed this data for measurement of soil quality of the different areas of Sundarbans. To assess the soil quality, six components have been used: Organic carbon, Bulk Density, EC, pH, N and P as shown in Table 7.2. Then we have tried to find out the relation among these components, which is known as correlation. Finally, we used hierarchical clustering to find the result.

Source:Field Survey: To assess soil quality, we have measured different soil nutrients which influence the soil quality of the mangrove forest. We have used statistical tools to measure the mean value of the soil nutrients present in different layers of our sample source.
(i) Organic Carbon: Carbon is important in retention of water, improving soil texture in soil [15]. From our study, we have found that the presence of organic carbon is 1.84% to 2.09%. The highest organic carbon exists in 15-30 cm soil depth.
(ii) Bulk Density is another important parameter to assess soil quality. In our study, bulk density in Sundarbans soil is 1.12 to 1.19 and the presence of bulk density is high in 15 to 50 cm depth.
(iii) Nitrogen has significant impact on determining soil quality.[16] found that nitrogen plays a critical role in retaining other organic particles in soil. The presence of nitrogen in the soil of our study is about 0.65 mg per gram. The second and third layers hold more nitrogen than other layers.
(iv) pH is an important element which has great influence in different chemical reactions in soil and supporting different types of fauna and trees in soil. Our study shows that the pH value of Sundarbans soil is 7.13 to 7.33 which means the soil is slightly acidic in nature.
(v) Phosphorous is important in soil to promote the growth of agriculture and forests [17]. Therefore, we measure the presence of phosphorous in the soil of Sundarbans. From our study, we can conclude that the phosphorous level of Sundarbans per mg is approximately 0.35 to 0.39 mg per gram.
(vi) Electrical Conductivity expresses the capacity of soil to transmit electricity. Our study shows that electrical conductivity in the soil of Sundarbans is 2.15, 3.86,4.32 and 5.61 in different layers of soil. Depth and electrical conductivity are positively correlated, which means that an increase in depth increases the electrical conductivity of soil.
(vii) Correlation among the Parameters: Correlation among the soil parameters are described in Table 7.3.

The table helps us show the correlation between different soil nutrients in terms of assessing soil quality. From our research, we have found that Nitrogen has positive correlation with Carbon. The coefficient of correlation is 0.806, as it is positive correlation, we can say if nitrogen increases, carbon will increase in the soil of Sundarbans. The relation between bulk density and Carbon is negative, with coefficient of $-.555$, which means the an increase in Carbon will simultaneously decrease the bulk density of the soil. The relation between nitrogen and Phosphorous is also positive and the correlation coefficient between them is 0.627. pH has no relation to the other parameters. It seems that it is constant or fixed in a different region, and from our study we found that most of the area belonging in our study has pH 7.15 to 7.3, which means that the pH level of the Sundarbans is slightly acidic.

Clustering: Clustering is used to show data with the same characteristics. The results from our experiments are described below. This dendrogram in Figure 7.4 shows the linkage between clusters. In Table 7.4, area of the same cluster bear out the same characteristics. Different area of cluster shows different characteristics.

Table 7.3 Correlation

Organic	Pearson Correlation	1	0.555**	.258*	0.009	0.806**	0.642**
	Sig. (2-Tailed)		0.000	0.047	0.943	0.000	0.000
	N	60	60	60	60	60	60
Bulk Density (G/Cm3)	Pearson Correlation	0.555**	1	0.032	0.079	0.477**	0.262*
	Sig. (2-Tailed)	0.000		0.810	0.548	0.000	0.043
	N	60	60	60	60	60	60
Ec (Ms/Cm)	Pearson Correlation	0.258*	0.032	1	0.147	0.124	0.220
	Sig. (2-Tailed)	0.047	0.810		0.262	0.344	0.091
	N	60	60	60	60	60	60
Ph	Pearson Correlation	0.009	0.079	0.147	1	0.202	0.022
	Sig. (2-Tailed)	0.943	0.548	0.262		0.123	0.869
	N	60	60	60	60	60	60
N (Mg/G)	Pearson Correlation	0.806**	0.477**	0.124	0.202	1	0.627**
	Sig. (2-Tailed)	0.000	0.000	0.344	0.123		0.000
	N	60	60	60	60	60	60
P (Mg/G)	Pearson Correlation	0.642**	0.262*	0.220	0.022	0.627**	1
	Sig. (2-Tailed)	0.000	0.043	0.091	0.869	0.000	
	N	60	60	60	60	60	60

** Correlation is Significant at 0.01 Level (2-Tailed)

The region which bears the similar characteristics with these soil nutrients of different region of Sundarbans. We have found 5 clusters using the Hierarchical Clustering algorithm. Harbaria S2(15-30), Akram Point S3(30-50), Akram Point S1(15-30), Sutarkhali S2(30-50), Harbaria S2(30-50), Sutarkhali S1(30-50), Sutarkhali S1(50-100), Akram Point S2(50-100), Harbaria S1(50-100), Sutarkhali S3(15-30), Harbaria S2(50-100), Sutarkhali S2(50-100), Sutarkhali S3(30-50), Karomjol S1(30-50), Karomjol S1(30-50) and Alorkol S3(50-100) have the same soil nutrient qualities and are situated in cluster 1. The regions of cluster 2 are Karomjol S3(30-50, Karomjol S3(50-100), Karomjol S1(50-100), Karomjol S1(15-30), Karomjol S3(15-30),

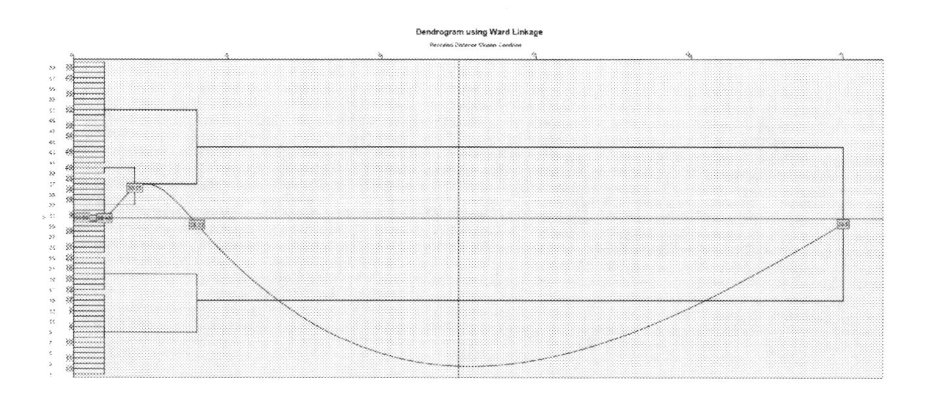

Figure 7.4 Dendrogram

Table 7.4 Interpretation of Clustering Result

Cluster	Region
Cluster 1	Alorkol S3(50-100) Harbaria S2(15-30) Akram Point S3(30-50) Akram Point S1(15-30) Sutarkhali S2(30-50) Harbaria S2(30-50) Sutarkhali S1(30-50) Sutarkhali S1(50-100) Akram Point S2(50-100) Harbaria S1(50-100 Sutarkhali S3(15-30) Harbaria S2(50-100) Sutarkhali S2(50-100) Sutarkhali S3(30-50) Karomjol S1(30-50) Karomjol S1(30-50)
Cluster 2	Akram Point S3(50-100) Sutarkhali S3(50-100) Karomjol S3(30-50 Karomjol S3(50-100) Karomjol S1(50-100) Karomjol S1(15-30) Karomjol S3(15-30)
Cluster 3	Harbaria S1(0-15) Harbaria S3(0-15) Harbaria S3(15-30) Harbaria S1(15-30) Karomjol S2(15-30) Akram Point S2(30-50)) Harbaria S1(30-50) Sutarkhali S3(0-15) Alorkol S1(30-50) Sutarkhali S2(0-15) Harbaria S2(0-15)) Sutarkhali S1(0-15) Karomjol S2(0-15) Karomjol S3(0-15) Alorkol S1(15-30)
Cluster 4	Alorkol S1(0-15) Alorkol S2(0-15)
Cluster 5	Sutarkhali S1(15-30)) Alorkol S2(50-100) Karomjol S2(30-50) Akram Point S2(15-30) Akram Point S2(0-15)) Akram Point S3(0-15) Harbaria S3(30-50) Alorkol S2(30-50) Akram Point S1(0-15) Alorkol S1(50-100) Akram Point S1(30-50) Karomjol S2(50-100) Alorkol S3(15-30) Harbaria S3(50-100) Alorkol S2(15-30) Alorkol S3(0-15) Alorkol S 1(30-50) Karomjol S1(0-15) Sutarkhali S2(15-30) Akram Point S3(15-30)

Akram Point S3(50-100), Sutarkhali S3(50-100) the similar characteristics. Harbaria S1(0-15), Harbaria S3(0-15), Harbaria S3(15-30), Harbaria S1(15-30), Karomjol S2(15-30), Akram Point S2(30-50), Harbaria S1(30-50), Sutarkhali S3(0-15), Alorkol S1(30-50), Sutarkhali S2(0-15), Harbaria S2(0-15), Sutarkhali S1(0-15), Karomjol

S2(0-15), Karomjol S3(0-15) and Alorkol S1(15-30) are in cluster 3 and have the same property. Cluster 4 the has the same quality and it has two regions of Sundarbans, Alorkol S1(0-15) and Alorkol S2(0-15). Cluster 5 includes Sutarkhali S1(15-30), Alorkol S2(50-100), Karomjol S2(30-50), Akram Point S2(15-30), Akram Point S2(0-15), Akram Point S3(0-15), Harbaria S3(30-50), Alorkol S2(30-50), Akram Point S1(0-15), Alorkol S1(50-100), Akram Point S1(30-50), Karomjol S2(50-100), Alorkol S3(15-30), Harbaria S3(50-100) , Alorkol S2(15-30), Karomjol S1(0-15), Sutarkhali S2(15-30), Akram Point S3(15-30), Alorkol S3(0-15), and Alorkol S1(30-50) also have similar characteristics.

7.5 CONCLUSION

Mangrove forests are huge natural resources with distinctive characteristics. The unique nature of their soil makes them superior to other forests. Our study shows that different amounts of nutrients in the Sundarbans soils exist in the proper amounts. Our study shows the presence of Organic Carbon is 1.84% to 2.25%, Bulk Density 1.12 to 1.21 (g/cm^3), Nitrogen 0.59 to 0.69 mg per gm, phosphorous 0.348 to 0.389 mg per gm, pH value is 7.13 to 7.19, electrical conductivity is 2.15 to 5.63 ms per cm in different layers of soil and which region carries similar characteristics depending on these soil nutrients. Despite its existence, day-by-day exploitation of Sundarbans is increasing. Our study will help policymakers get actual knowledge of Sundarbans soil quality, which will help make sufficient efforts to preserve this valuable natural resource. Further, more study is needed to find more viable results.

Bibliography

[1] Akhter, M. (2006). Remote sensing for developing an operational monitoring scheme for the Sundarban Reserved Forest, Bangladesh lengl.g.

[2] Clark, M. W., McConchie, D., Lewis, D. W., & Saenger, P. (1998). Redox stratification and heavy metal partitioning in Avicennia-dominated mangrove sediments: A geochemical model. *Chemical Geology*, 149(3-4), 147–171.

[3] Dabgerwal, D. K., & Tripathi, S. K. (2016). Assessment of surface water quality using hierarchical cluster analysis. *International Journal of Environment*, 5(1), 32–44.

[4] Duke, Norman C., Jan-Olaf Meynecke, Sabine Dittmann, Aaron M. Ellison, Klaus Anger, Uta Berger, Stefano Cannicci et al. A world without mangroves? *Science*, 317(5834), 41–42.

[5] FAO (1997) *State of the World's Forests, 1997.* Food and Agriculture Organization of the United Nations, Rome.

[6] Hasan, M. S. (2012). Sundarban Mangrove Forest from 1989 to 2009.

[7] Holtan, H., Kamp-Nielsen, L., & Stuanes, A. O. (1988). Phosphorus in soil, water and sediment: An overview. Phosphorus in freshwater ecosystems, 19–34.

[8] Hossain, M. D., & Nuruddin, A. A. (2016). Soil and mangrove: A review. *Journal of Environmental Science and Technology*, 9(2), 198–207.

[9] Kristensen, E. (2000). Organic matter diagenesis at the oxic/anoxic interface in coastal marine sediments, with emphasis on the role of burrowing animals. *Life at Interfaces and under Extreme Conditions*, 1–24.

[10] Norušis, M. J. (2009). *PASW Statistics 18 Statistical Procedures Companion*. Prentice Hall.

[11] Singh, S. K., Srivastava, P. K., Gupta, M., Thakur, J. K., & Mukherjee, S. (2014). Appraisal of land use/land cover of mangrove forest ecosystem using support vector machine. *Environmental Earth Sciences*, 71(5), 2245–2255.

[12] Richards, D. R., Thompson, B. S., & Wijedasa, L. (2020). Quantifying net loss of global mangrove carbon stocks from 20 years of land cover change. *Nature Communications*, 11(1), 1–7.

[13] Rawls, W. J., Pachepsky, Y. A., Ritchie, J. C., Sobecki, T. M., & Bloodworth, H. (2003). Effect of soil organic carbon on soil water retention. *Geoderma*, 116(1-2), 61–76.

[14] Sanderman, J., Hengl, T., Fiske, G., Solvik, K., Adame, M. F., Benson, L., ... & Landis, E. (2018). A global map of mangrove forest soil carbon at 30 m spatial resolution. *Environmental Research Letters*, 13(5), 055002.

[15] Saleh, A., & Belal, A. (2014, February). Delineation of site-specific management zones by fuzzy clustering of soil and topographic attributes: A case study of East Nile Delta, Egypt. In *IOP Conference Series: Earth and Environmental Science* (Vol. 18, No. 1, p. 012046). IOP Publishing.

[16] Schimel, D. S., Braswell, B. H., & Parton, W. J. (1997). Equilibration of the terrestrial water, nitrogen, and carbon cycles. *Proceedings of the National Academy of Sciences*, 94(16), 8280–8283.

[17] Tripathi, R., Shukla, A. K., Shahid, M., Nayak, D., Puree, C., Mohanty, S., ... & Nayak, A. K. (2016). Soil quality in mangrove ecosystem deteriorates due to rice cultivation. *Ecological Engineering*, 90, 163–169.

CHAPTER 8

A Machine Learning Approach to Clinically Diagnose Human Pyrexia Cases

Dipon Talukder

Chittagong University of Engineering and Technology, Chittagong, Bangladesh

Md. Mokammel Haque

Chittagong University of Engineering and Technology, Chittagong, Bangladesh

CONTENTS

For the past decade, a lot of research has been conducted to employ the ability of machine learning in the healthcare system to diagnose patients with diseases. It has been observed that incorporating machine learning into a modern diagnostic method will significantly reduce the chance of misdiagnosing outcomes. To have an impact on the healthcare system, this research provides a technique and a number of features for diagnosing pyrexia cases and presents experiments with the three most closely symptomized infectious diseases: Dengue, Malaria, and Typhoid. In order to develop the diagnostic model, we propose a set of 29 features for correctly classifying pyrexia cases. Furthermore, a dataset with the proposed features has been developed, consisting of symptoms of pyrexia patients, obtained from pyrexia patients admitted to the hospitals in five districts of Bangladesh. In addition, the collected data is analyzed in detail and the key findings of the study are discussed. The dataset is used to evaluate the accuracy of various machine learning algorithms. Despite the fact that many studies have found Support Vector Machine and Rough Set Theory to be the most efficient models in such classification tasks, the research found and recommends logistic regression with the proposed feature set to be the most effective, with a classification accuracy of 95.1%.

8.1 INTRODUCTION

Pyrexia is a condition where body temperature accelerates in an abnormal manner. Pyrexia is also commonly known as fever. Mild escalation of body temperature can be resolved by taking available counter medications such as Ibuprofen and Paracetamol. But a sudden high rise in body temperature can be life-threatening. Although there is variance among expertise about the right body temperature for humans, the value ranges from 99.0 to 100.9 degrees Fahrenheit [1, 2, 3]. Pyrexia is considered to be one of the most regular medical signs. It occurs in up to 75 percent of chronically ill adults [4]. Other than adults, pyrexia accounts for about 30 percent of children's visits to the doctor [5].

Among many explanations, Typhoid, Dengue, and Malaria are regarded as common causes of pyrexia [6]. In 2015 alone, 12.5 million people were infected globally by Typhoid [7]. The same year, 1.49M deaths were reported worldwide due to Typhoid [8]. Without treatment, the probability of mortality may be as high as 20% in this disease [9]. On the other hand, in the case of dengue, around 390 million new cases are registered every year, among which, approximately 40 thousand people die [10, 11]. Malaria caused 228 million cases worldwide in 2018, with approximately 405,000 deaths [12]. These infectious diseases are often misdiagnosed [13].

Between numerous applications, machine learning technology has shown promising results in the development of the healthcare system for the past few years [14, 16, 15]. Machine Learning has the ability to find a hidden pattern among the complex medical data that is creating an impact on clinical diagnosis systems.

This work was undertaken to achieve a specific set of goals. The main objective of this work is to create an impact in the healthcare system through the power of machine learning and to contribute to the machine learning society for the further development of the clinical system.

The main contributions of this work can be quantified as follows:

- The research proposes a set of 29 features for diagnosing pyrexia cases. Due to lack of data, the experiment was conducted to classify only three closely symptomized infectious pyrexia diseases(Dengue, Malaria, Typhoid).
- A dataset comprising symptoms of pyrexia patients was developed from patients admitted to hospitals in Bangladesh.
- Analysis of data that is obtained from the hospitals.

- Machine learning models are evaluated using the developed dataset and based on the evaluation result, the research recommends an appropriate model for the task.

8.2 RELATED HEALTHCARE RESEARCH

Practitioners can detect diseases by looking at physiological evidence, environmental conditions, and genetic factors. Machine learning enables the development of models that link a wide variety of features to a disease. In this study, Oguntimilehin et al. [17] diagnosed five classes of typhoid using a machine learning approach. Data used to train the model has been collected from Typhoid patients in Nigeria. Symptoms are collected and labeled by medical experts. In the literature, authors have quoted medical experts who have suggested that although typhoid shares the same symptoms as malaria and dengue, they are still differentiable with a large number of feature variables. The researchers used a rule-based RST algorithm and defined 18 rules to classify the level of Typhoid. In the literature, it is shown that researchers were able to achieve 96% accuracy in the test dataset. But the work holds a huge drawback. Symptoms of the patients are collected from only one hospital, hence the dataset contains the typhoid symptoms of only a small region, which caused a lack of variance in the dataset.

In another study, researchers diagnosed the severity of dengue fever using a machine learning approach [18]. In this case, researchers used an SVM algorithm to locate the optimal loci classification subset and used a Neural Network for the classification of acute dengue fever or moderate dengue fever. For this purpose, the authors used human genome data instead of symptomatic data of the dengue patients. The work also claimed that the proposed methodology can be applied to diagnose any genetically influenced disease.

In a paper, experimenters proposed a novel method for classifying dengue fever using an online learning method [19]. The proposed system initially trained the model with a few training samples, but it is able to gain experience with time through the online learning method. Initial training, which is referred to in the literature as offline training, is done using SVM and the Random Forest algorithm. The online training is done using the AROW and CWL algorithms. The researchers achieved 98% accuracy for predicting dengue patients. Thirteen features are selected for the ML model. However, the system requires the model to integrate with healthcare system software for collecting symptoms of patients for the online learning process. Also, no validation is done before labeling the data instances before training.

In a survey, various machine learning algorithms were used to diagnose various diseases [20]. The experimenters analyzed and recommended appropriate algorithms for particular disease detection. The analysis showed the Rough Set Theory algorithm to be most effective for predicting infectious diseases such as dengue.

Researchers also implemented a machine learning approach to classify acute and non-acute Covid 19 patients [21]. The scientists have carefully chosen 26 features from the symptom list of covid patients. Symptoms of the patients are recorded from

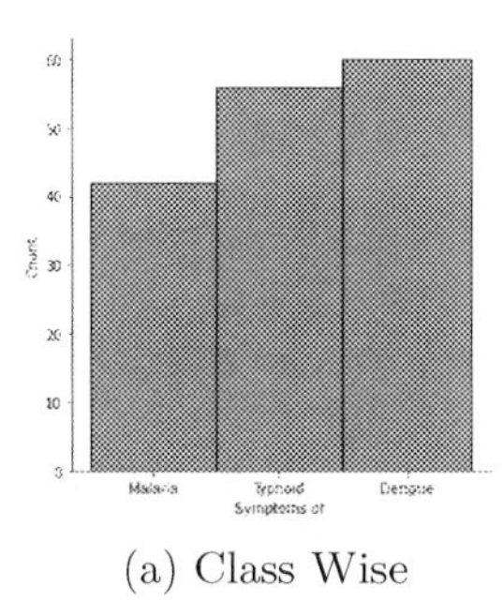

(a) Class Wise

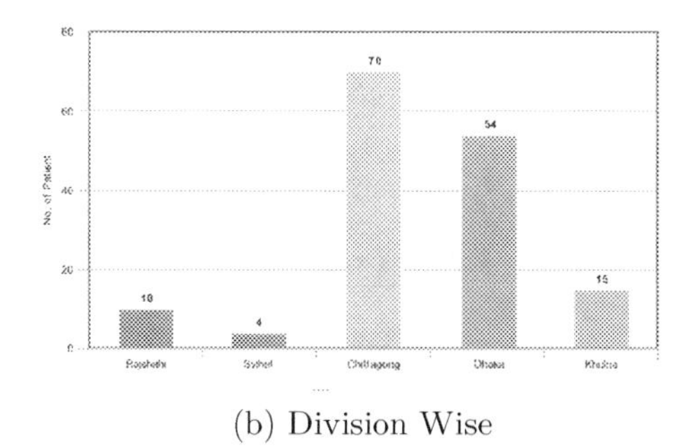

(b) Division Wise

Figure 8.1 Number of Records

Wuhan. Researchers found the Random Forest Algorithm provided the best accuracy in this regard and pulled 90% accuracy in classification results.

Researchers developed a model to diagnose malaria using patient information [22]. The researchers have compared the developed dataset using six different machine learning models. According to their study, random forest has proved to be the best model in this regard. The authors also concluded that machine learning technology can be successfully applied to predict malaria using information gathered from the patient.

Experimenters from another paper, recommended a machine learning approach to predict and provide treatment of Typhoid cases [23]. The authors implemented a decision tree algorithm for the task and achieved 95% accuracy over the test dataset.

Another research paper [24], proposed a hybrid system to classify disease caused by infection. The authors originally classified two infectious diseases: Dengue and Typhoid. The collected dataset has been fed through many traditional machine learning algorithms to compare the result. Accuracy of 97.2% has been achieved through their work.

8.3 DATASET DESCRIPTION

This section discusses and provides an overview of the obtained dataset.

8.3.1 Dataset Collection

The dataset contains symptoms of 153 patients admitted at the hospitals and diagnosed with either Dengue, Typhoid or Malaria, collected between September 2019 and December 2019. The dataset was collected in accordance with the 29 features, designed for the purpose. Feature design is described and explained in the next section. As per feature design, each instance comprises 29 features to efficiently classify three closely symptomized infectious diseases.

Fig 8.1(a) represents the distribution of 153 data records according to each class. The dataset is slightly unbalanced, which can be negligible in this case.

Data was recorded and collected from various divisions of Bangladesh. Since it was collected from a wide geographical region, the dataset contains a good amount of variance in this perspective. Figure 8.1(b) shows the number of patient records

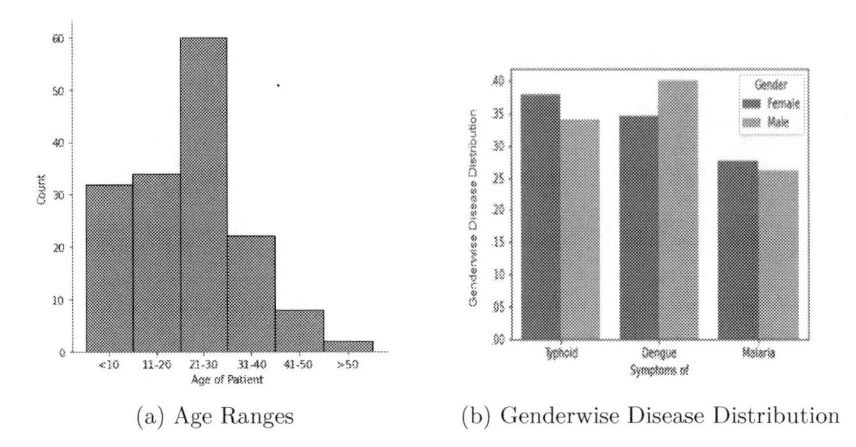

(a) Age Ranges (b) Genderwise Disease Distribution

Figure 8.2 Distribution Overview based on Age and Gender

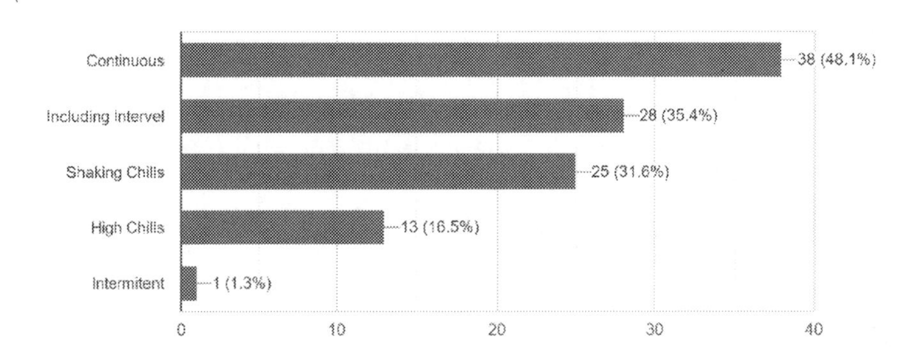

Figure 8.3 Fever Characteristics

collected from each division. Most of the patient records were registered from Chittagong, which includes hill districts like Rangamati, Bandarban and Khagrachori. The diseases are most common in the Chittagong division due to the presence of hill districts.

8.3.2 Data Analysis and Deductions

Symptoms were collected from patients of various ages. Figure 8.2(a) plots the number of patients against the age range of the patients present in the dataset. The dataset has the highest number of recorded symptoms of the admitted patients 21 to 30 years of age. For further exploration, in Figure 8.2(b) we have plotted the proportion of the patients who have been diagnosed with dengue, malaria or typhoid grouped by gender to gain more insight.

In Figure 8.3, we have represented the fever characteristics of the patients in terms of percentage. The type of fever that a pyrexia patient mostly experiences can be observed in this figure.

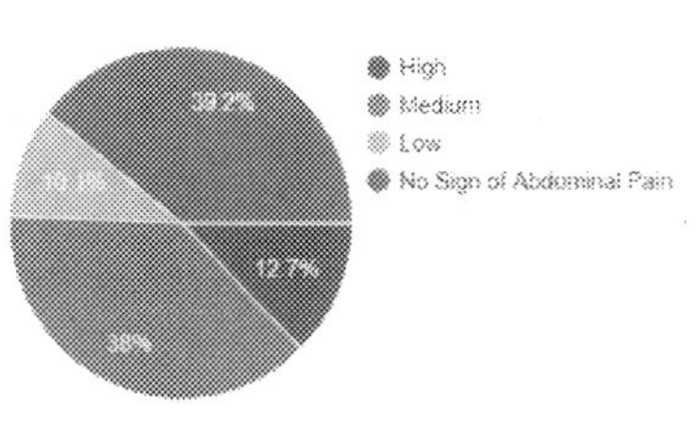

(a) Abdominal Pain Pie Chart

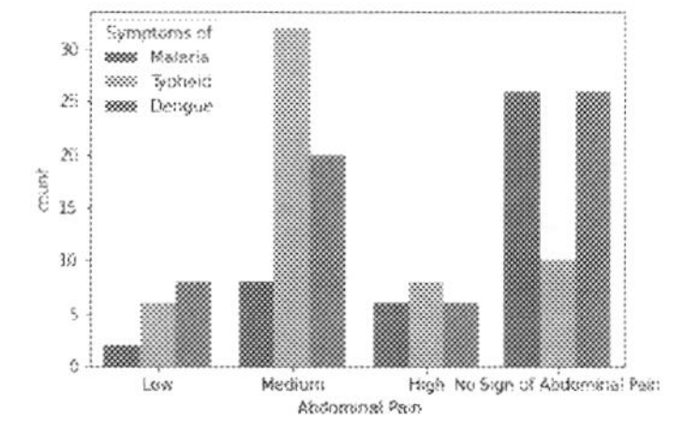

(b) Disease Wise Distribution

Figure 8.4 Exploration of Feature Abdominal Pain

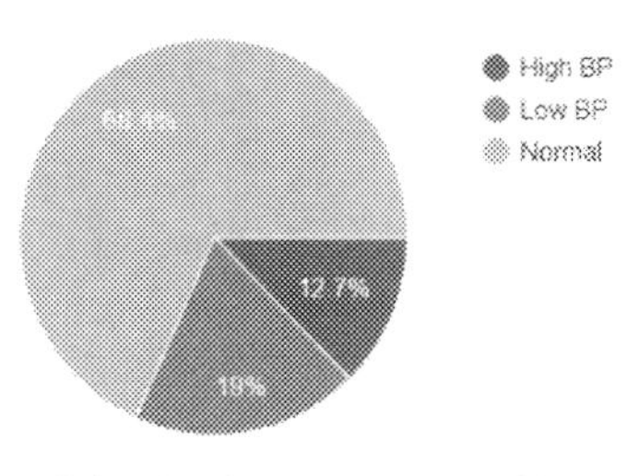

(a) Blood Pressure Pie Chart

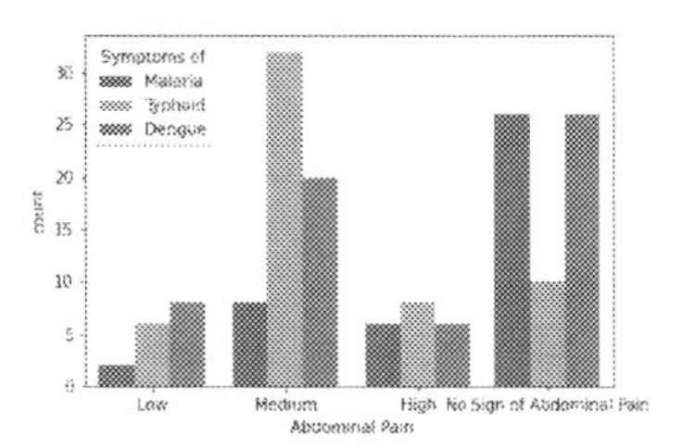

(b) Disease Wise Distribution

Figure 8.5 Exploration of Feature Blood Pressure

Figure 8.4 explores one of the features used for the machine learning model. In subfigure 8.4(a) the feature abdominal-pain pie chart is plotted showing that among all patients, 39.2% have not experienced abdominal pain as a disease symptom. On the other hand, it can be deduced from subfigure 8.4(b) that patients who have not experienced abdominal pain are more likely to be diagnosed with either Malaria or Dengue.

Figure 8.5 explores another one of the core features used for the machine learning model. In subfigure 8.5(a), the blood pressure pie chart is plotted showing that among all patients, 68.4% have not experienced blood pressure as a disease symptom.

This is not surprising given that the majority of the patients were between the ages of 21 and 30 as seen before. On the other hand, subfigure 8.5(b) shows the number of patients who experienced or did not experience blood pressure grouped by the disease diagnosed.

8.4 FEATURE SELECTION

This section discusses and explains the proposed features used for the experiment.

8.4.1 Primary Feature Selection

The primary feature selection process is the most crucial part of our research. In this research, the primary features are the symptoms of pyrexia patients. Previously, researchers have designed and proposed different kinds of feature-sets that have to be accounted for to diagnose pyrexia patients. But to differentiate three diseases with the most similar symptoms, such as Dengue, Malaria, Typhoid, a large number of features with a cautious selection process is needed as the medical expert suggested [17]. For this, we propose a set of 29 features to classify pyrexia diseases. The features are listed below:

- Age of Patient
- Gender
- Diabetes Condition
- Blood Pressure
- Fever Type
- Fever Characteristics
- Headache
- Joint Pain
- Muscle Pain
- Fatigue (Tiredness)
- Nausea
- Vomiting
- Sensitivity to Light
- Insomnia
- Abdominal Pain
- Skin Rash
- Stuffy / Runny Nose
- Cough
- Diarrhea
- Sweating
- Constipation
- Dizziness
- Sore Throat
- Back Pain
- Chest Congestion
- Eye Pain
- Poor Appetite
- Convulsions
- Bleeding

These features were chosen and designed with the assistance of medical professionals and the book *Davidson's Principles and Practice of Medicine* [25].

8.4.2 Final Feature Selection

In Figure 8.6 the correlation matrix of the features is represented. The Pearson Correlation Coefficient technique has been used to compute the pairwise correlation of the selected primary features and to generate the correlation matrix in Fig. 8.6.

We can analyze from the correlation matrix the topmost key factors that have a higher relationship with the target variable. The key factors along with their relation strength with the output is analyzed and the surprising outcome is observed in Table 8.1.

After examination, features like Eye Pain, Cough, Diarrhea, and Vomiting are found to be strongly positively correlated with the outcome variable with the Coefficient values of 0.72, 0.39, 0.39, 0.38. On the other hand, features like Insomnia, Light Sensitivity and Patient Age are strongly negatively correlated with the output variable with the coefficient values of −0.76, −0.43, −0.36. Moreover, the Nausea, Gender, Headache, Blood Pressure, and Poor Appetite features contribute much less with the coefficient value nearly zero.

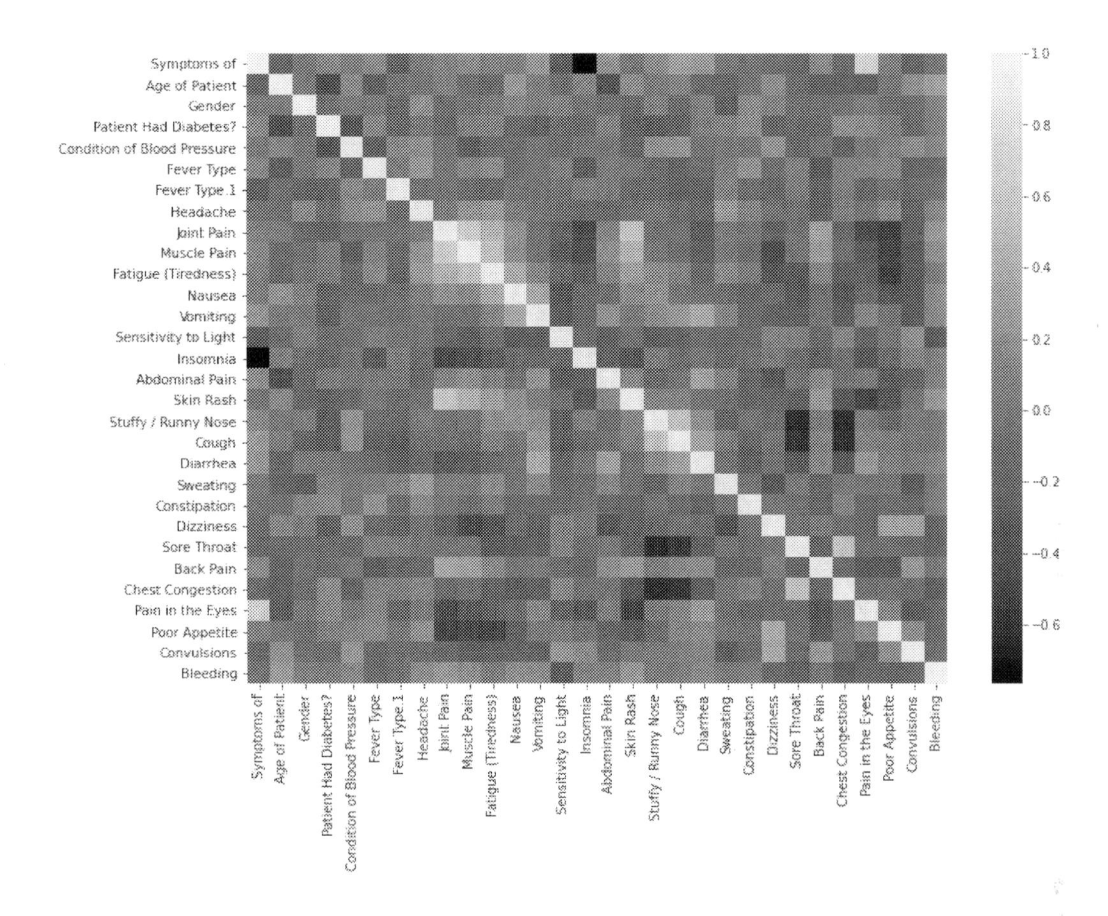

Figure 8.6 Correlation Matrix

Table 8.1 Key Factors

Feature	Relation Strength Factor
Insomnia	0.76
Eye Pain	0.71
Cough	0.39

However, in our research work, we did not remove any of the features since we did not want to lose any of the patient information in these instances, regardless of their significance.

8.5 MODEL EVALUATION

In our work, to compare the performance of machine learning models against our collected and preprocessed dataset, we have used the six most popular machine learning classifiers and trained them. Each algorithm implementation was imported from scikit-learn, the most popular machine learning library. The parameter values for each method are described in detail in Appendix A. For the evaluation process and to

Table 8.2 Performance Evaluation of ML Algorithms

Classifier	Accuracy	Precision	Recall	F1 Score
SVM	82.5%	0.84	0.82	0.84
Gaussian Naive Bayes	93%	0.92	0.93	0.92
K-Neighbors	75.1%	0.74	0.74	0.74
Decision Tree	82.5%	0.90	0.83	086
Random Forest	85%	0.925	0.923	0.923
Logistic Regression	95.1%	0.94	0.94	0.94

determine the best-performing, model we have calculated the model accuracy, precision, recall and F1 score of each trained model. For high-quality performance evaluation, we have split the dataset into 75% and 25% for training and testing respectively. Each of the models was trained with the shuffled 75% of data instances and tested with the other 25%.

As we can analyze Table 8.2, the logistic regression algorithm provides the best performance in this case with an accuracy of 95.1%. The algorithm also presented promising results with high precision, recall and F1 score.

In previous research work, [24] implemented a hybrid solution to diagnose only typhoid and dengue, where in this work a wide range of features are used and able to diagnose three infectious diseases with high accuracy. Also, in healthcare research, [21] designed 26 features to only classify the pyrexia severity level of another infectious disease, COVID-19, and obtained 90% accuracy, compared to which, our work diagnosed three infectious diseases with higher accuracy with our designed 29 features. Similar works have also been conducted where only the severity of typhoid disease [17] and dengue disease [18] was predicted. Also, where various surveys [20] suggested an RST SVM algorithm for this kind of disease diagnosis, our study found logistic regression to be the best model in this case.

8.6 RESULT ANALYSIS

In this section, we discuss the performance of the logistic regression classifier on our test dataset and try to visualize the reason for logistic regression to be the best algorithm in this regard.

In Fig 8.7 a confusion matrix of the test cases is plotted for error analysis purposes. Here, class 0, class 1, and class 2 represent Dengue, Malaria, and Typhoid respectively. As shown in Fig 8.7, out of 40 test cases, only 2 of the cases were misdiagnosed. One Dengue case is predicted as Malaria and the other Malaria case was predicted as Dengue. Considering the fact that diseases are closely symptomized, the number of false predictions is relatively very low.

Logistic Regression is a statistical method and a linear classifier for binary classification. For multiclass classification, logistic regression adopts the One vs All Method to classify each class. It is impractical to visualize logistic regression performance for 29-dimensional data. However, to understand the classification, the 29-dimension data

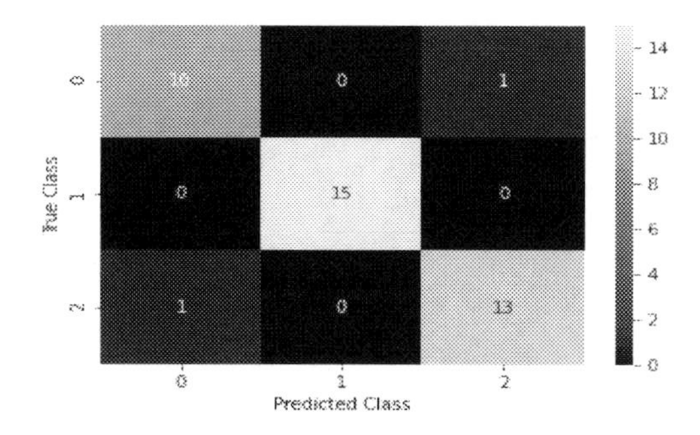

Figure 8.7 Confusion Matrix (Logistic Regression)

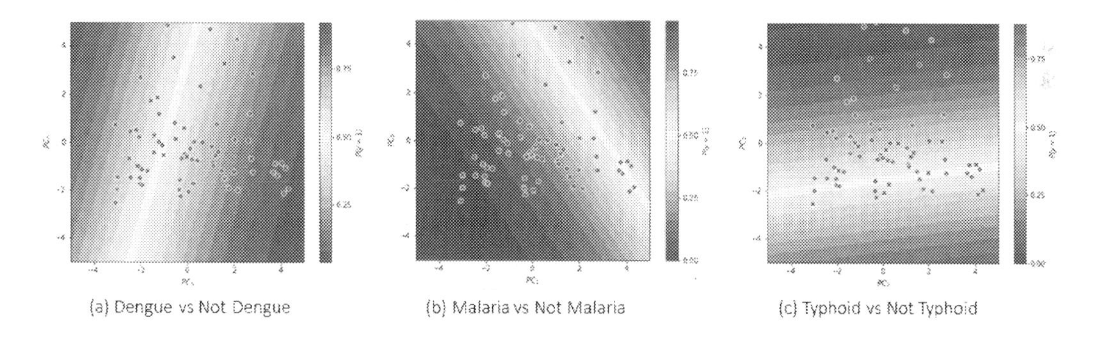

Figure 8.8 Logistic Regression Model Performance Analysis

is reduced to 2 dimensions using the compression technique, Principal Component Analysis. In Fig. 8.8(a) we plotted Principle-Component-1 and Principle-Component-2 with a logistic regression decision boundary to observe the internal performance of the model to classify Dengue and Not Dengue cases. In this case, Malaria and Typhoid cases are considered the same, the Not Dengue class. Similarly, in Figs. 8.8(b) and 8.8(c) we plotted the logistic regression decision boundary for diagnosing Malaria and Typhoid Cases. Examining Fig. 8.8 it can be said that although the figure is useful to envision the machine learning model functioning, there are several false prediction cases seen in each subfigure. Though blue points should be in the blue region and red points should be in the red region, around 28% of the data points are misplaced and residing on the wrong side of the white decision boundary line in all the subfigures. The reason can be discussed using Table 8.3.

Explained Variance Ratio of the first 15 principal components are calculated using eigenvectors and eigenvalues and listed in Table 8.3. From the table, the first two principal components retain barely 30% of the original information, losing over 70%. With conserving only 30% of the information, the logistic regression model classified the data admirably in Fig. 8.8 causing the effective performance of logistic regression in our dataset. To obtain and visualize a good result, at least 80% of the information has to be reserved [26]. In that case, at least the first 13 principal components have to be plotted, which is not possible to visualize in this case. This experimental outcome verifies another analysis result that we observed earlier from the correlation matrix,

Table 8.3 Variance Ratio of Principal Components

Principal Component	Explained Variance Ratio (EVR)	Cumulative EVR
PC-1	0.141	0.141
PC-2	0.135	0.276
PC-3	0.095	0.371
PC-4	0.064	0.435
PC-5	0.061	0.496
PC-6	0.059	0.555
PC-7	0.052	0.607
PC-8	0.045	0.652
PC-9	0.043	0.695
PC-10	0.034	0.729
PC-11	0.033	0.762
PC-12	0.027	0.789
PC-13	0.025	0.814
PC-14	0.024	0.838
PC-15	0.021	0.859

that all features are densely correlated, which was the very reason that none of the features were dropped or modified before training the machine learning model.

8.7 CONCLUSION AND FUTURE WORKS

In our work, we tried to create an impact on the modern healthcare system by providing a methodology to diagnose human pyrexia diseases. We may conclude from the experiment results and model accuracy that the proposed feature set is more effective in classifying pyrexia cases. Despite the fact that the features were developed to detect at least 5 pyrexia patients, the other two cases were dropped due to lack of data. Chikungunya and Swine flu were the other two pyrexia cases. As it is already assumed, the most important future work is expanding the dataset and adding other pyrexia case information. The dataset only contains details on 153 patients. Although the number of patients in the previous studies ranged from 200 to 300, We conclude that at least 500 cases (150 per disease) are needed to analyze more accurately.

ACKNOWLEDGEMENT

The research acknowledges and expresses tremendous gratitude to all individuals without whom the work would not be possible to finish. We cannot express enough thanks to Dr. Puspen Das Gupta, Dr. Rubiath Farhin Etu, Dr. Mejbah Uddin, Dr. Diptanil Das, Dr. Joy Deb, Dr. Md. Hossain Al Imran and many more medical experts have helped to develop the dataset by collecting symptoms from patients and also for providing expert support with the work review.

Bibliography

[1] Jameson JL, Fauci AS, Kasper DL, et al. Harrison's Principles of Internal Medicine. McGrawHill Education; 2018.

[2] Axelrod YK, Diringer MN. Temperature management in acute neurologic disorders. *Neurologic Clinics*. 2008;26(2):585–603.

[3] Laupland KB. Fever in the critically ill medical patient. *Critical Care Medicine*. 2009;37(7):S273–S278

[4] Kiekkas P, Aretha D, Bakalis N, et al. Fever effects and treatment in critical care: Literature review. *Australian Critical Care*. 2013;26(3):130–135.

[5] Sullivan JE, Farrar HC, et al. Fever and antipyretic use in children. *Pediatrics*. 2011;127(3):580–587.

[6] 6 most common diseases during monsoon: Dengue, malaria, typhoid, cholera and more- symptoms and prevention; https://www.timesnownews.com/health/article/6-most-common-diseases-during-monsoon-dengue-malaria-typhoid-cholera-and-more-symptomsand-prevention/456319.

[7] Vos, T., Allen, C., Arora, M., Barber, R., Bhutta, Z., Brown, A., Carter, A., Casey, D., Charlson, F., Chen, A., et al. (2016). Global, regional, and national incidence, prevalence, and years lived with disability for 310 diseases and injuries, 1990–2015: A systematic analysis for the Global Burden of Disease Study 2015. *The Lancet*, 388(10053), 1545–1602.

[8] Wang, H., Naghavi, M., Allen, C., Barber, R., Bhutta, Z., Carter, A., Casey, D., Charlson, F., Chen, A., Coates, M., & others (2016). Global, regional, and national life expectancy, all-cause mortality, and cause-specific mortality for 249 causes of death, 1980–2015: A systematic analysis for the Global Burden of Disease Study 2015. *The Lancet*, 388(10053), 1459–1544.

[9] World Health Organization, et al. (2019). Typhoid vaccines: WHO position paper, March 2018–Recommendations. *Vaccine*, 37(2), 214–216.

[10] World Health Organization. Dengue and severe dengue. World Health Organization. Regional Office for the Eastern Mediterranean; 2014.

[11] Roth, G., Abate, D., Abate, K., Abay, S., Abbafati, C., Abbasi, N., Abbastabar, H., Abd-Allah, F., Abdela, J., Abdelalim, A., et al. (2018). Global, regional, and national age-sex-specific mortality for 282 causes of death in 195 countries and territories, 1980–2017: A systematic analysis for the Global Burden of Disease Study 2017. *The Lancet*, 392(10159), 1736–1788.

[12] World Health Organization. (2019). *World Malaria Report 2019*. World Health Organization.

[13] Waggoner, J., Gresh, L., Vargas, M., Ballesteros, G., Tellez, Y., Soda, K., Sahoo, M., Nuñez, A., Balmaseda, A., Harris, E., et al. (2016). Viremia and clinical presentation in Nicaraguan patients infected with Zika virus, chikungunya virus, and dengue virus. *Clinical Infectious Diseases*, ciw589.

[14] Bhardwaj, R., Nambiar, A., & Dutta, D. (2017). A study of machine learning in healthcare. In *2017 IEEE 41st Annual Computer Software and Applications Conference* (COMPSAC) (pp. 236–241).

[15] Wiens, J., & Shenoy, E. (2018). Machine learning for healthcare: on the verge of a major shift in healthcare epidemiology. *Clinical Infectious Diseases*, 66(1), 149–153.

[16] Callahan, A., & Shah, N. (2017). Machine learning in healthcare. In *Key Advances in Clinical Informatics* (pp. 279–291). Elsevier.

[17] Oguntimilehin, A., Adetunmbi, A., & Abiola, O. (2013). A Machine learning approach to clinical diagnosis of typhoid fever. *A Machine Learning Approach to Clinical Diagnosis of Typhoid Fever*, 2(4), 1–6.

[18] Davi, C., Pastor, A., Oliveira, T., Lima Neto, F., Braga-Neto, U., Bigham, A., Bamshad, M., Marques, E., & Acioli-Santos, B. (2019). Severe dengue prognosis using human genome data and machine learning. *IEEE Transactions on Biomedical Engineering*, 66(10), 2861–2868.

[19] Srivastava, S., Soman, S., Rai, A., & Cheema, A. (2020). An online learning approach for dengue fever classification. In *2020 IEEE 33rd International Symposium on Computer-Based Medical Systems* (CBMS) (pp. 163–168).

[20] Fatima, M., Pasha, M., et al. (2017). Survey of machine learning algorithms for disease diagnostic. *Journal of Intelligent Learning Systems and Applications*, 9(01), 1.

[21] Chen, Y., Ouyang, L., Bao, F., Li, Q., Han, L., Zhu, B., Ge, Y., Robinson, P., Xu, M., Liu, J., & others (2020). An interpretable machine learning framework for accurate severe vs non-severe Covid-19 clinical type classification. Available at SSRN 3638427.

[22] Lee, Y., Choi, J., & Shin, E.H. (2021). Machine learning model for predicting malaria using clinical information. *Computers in Biology and Medicine*, 129, 104151.

[23] Oguntimilehin, A., Adetunmbi, A., & Olatunji, K. (2014). A machine learning based clinical decision support system for diagnosis and treatment of typhoid fever. *A Machine Learning Based Clinical Decision Support System for Diagnosis and Treatment of Typhoid Fever*, 4(6), 1–9.

[24] Sajana, T., Syamala, M., Maguluri, L., & Kumari, C. (2021). A hybrid approach for classification of infectious diseases. *Materials Today: Proceedings*.

[25] Ralston, S., Strachan, M., Britton, R., Penman, I., & Hobson, R. (2018). Davidsons Principles & Practice of Medicine. Elsevier.

[26] Lindgren, I.. (2020). Dealing with Highly Dimensional Data using Principal Component Analysis(PCA); https://towardsdatascience.com/dealing-with-highly-dimensional-datausing-principal-component-analysis-pca-fea1ca817fe6#:Ÿ:text=Theexplainedvarianceratiois,or80%toavoidoverfitting

CHAPTER 9

Prediction of the Dengue Incidence in Bangladesh Using Machine Learning

Md. Al Mamun
Rajshahi University of Engineering & Technology, Rajshahi, Bangladesh

Abu Zahid Bin Aziz
Rajshahi University of Engineering & Technology, Rajshahi-6204, Bangladesh

Md. Palash Uddin
Deakin University, Australia

Md Rahat Hossain
Central Queensland University, North Rockhampton, QLD, Australia

CONTENTS

ATHOUGH the organisms or insects are normally harmless, but under certain conditions, some may cause serious diseases. Dengue is such a disease; it is transmitted by mosquitoes, caused by any of the four related dengue viruses, and causes infections leading to fever and fatigue. Taking rest and home remedies can sometimes cure mild infections while sometimes life-threatening infections may need hospitalization. As such, an outbreak of dengue is one of the top diseases causing the most deaths worldwide including in Bangladesh. Since 1964, Bangladesh has experienced the sporadic occurrence of dengue until 2000 when the first epidemic of dengue was reported in the capital city, Dhaka. Since then, the disease has shown an annual occurrence in all major cities of the country. The state-of-the-art methodologies, e.g., machine learning approaches, are now being used in many countries for early predicting or forecasting dengue cases. In this chapter, we propose machine learning algorithms for the early prediction of dengue cases in Bangladesh. We collect and preprocess meteorological data of Dhaka city to fit the machine learning models. In this work, we propose a weighted average ensemble technique of five machine learning methodologies for predicting the number of dengue cases per month from meteorological data. The proposed approach produced promising results in predicting the dengue cases for our testing data samples, which shows great potential for employing machine learning algorithms successfully for early dengue incident prediction in various cities of Bangladesh. We hope that our methodology can contribute to further research on predicting dengue incidents in Bangladesh using meteorological data.

9.1 INTRODUCTION

The outbreak of dengue is one of the top diseases causing the most deaths worldwide. According to the World Health Organization, dengue infection has increased 30-fold globally over the past five decades and about 50–100 million new infections occur annually in more than 80 countries [1]. At present, the most outstanding and prominent steps to analyze the situation are (i) finding the available data sources; (ii) evaluating data preparation techniques; (iii) representing data; (iv) training and testing of forecasting models and methods; (v) dengue forecasting models evaluation approaches; and (vi) finally, proposing future challenges and possibilities in forecasting modeling of dengue outbreaks. It is obvious that the training and testing of forecasting models and methods (stage-iv) make up the core functional block that guides us to design a proper remedy to lessen the dengue outbreaks. Dengue fever had been intermittent in Bangladesh since 1964, before the first outbreak was recorded in the capital city of Dhaka in 2000. Since then, the disease has resurfaced every year in all the country's major cities.

Nowadays, many countries are using cutting-edge methodologies, such as machine learning methods, to predict or forecast dengue cases early, so that immediate actions could be taken for effective remedy. From this motivation, we suggest using machine learning algorithms to predict dengue fever early in Bangladesh by collecting and preprocessing the prior dengue data. The main objectives of the chapter are as follows.

- Collecting dengue data of the past 15 years and preparing a well-established dengue database for Bangladesh, which can be a vital resource for further study.
- Applying machine learning algorithms, such as linear regression, XgBoost, and Support Vector Machine (SVM) for dengue prediction (datasets from other countries are used as well). The models are used for the future prediction of dengue in Bangladesh. The main purpose of choosing these algorithms is that the recent works on this type of study are going on using these algorithms.

Besides, linear regression makes the estimation procedure simple and it is easy to implement. Xgboost has fast execution and SVM is more effective in high-dimensional spaces.

- Modification of models to get better performance. Having those advantages, we work on those algorithms for effective modification.

To this end, we aim for (i) dataset preparation (data collection, data cleaning, feature reduction etc.); (ii) analysis of the machine learning algorithms (linear regression, XgBoost, SVM) for dengue prediction; (iii) analysis of dengue outbreaks; and (iv) mitigating this dengue fever epidemic.

We organize the following part of the chapter as follows. Section 9.4 presents the state-of-the-art works for predicting and forecasting dengue cases worldwide. In Section 9.3, we describe the proposed workflow for predicting and forecasting dengue incidents in Bangladesh. Section 9.4 provides our experiments and result analyses and finally, we conclude our findings and observations in Section 9.5.

9.2 LITERATURE REVIEW

Early prediction of dengue cases dates back several years. In [2], a machine learning approach was proposed for the prediction of dengue fever severity solely based on human genome data. A novel multi-stage machine learning approach combination of auto-encoding, window-based data representation, and trend-based temporal clustering to predict dengue incidence in Mexico was proposed in [3]. Therefore, some scattered works are now in progress as dengue is now a big concern of the south Asian region. In [4], the possibility of applying machine learning and deep learning approaches to predict the number of confirmed dengue fever cases in Kuala Lumpur was proposed. Furthermore, in [5], the authors applied machine learning algorithms on clinical features such as blood profile and vitals to predict dengue fever. Environmental data, epidemiological, and socioeconomic data have been utilized. The authors proposed that the genetic algorithm enhanced recurrent neural network (GA-RNN) performs better prediction than the linear regression and decision tree. In [6], epidemiological and meteorological data were analyzed for a two-step prediction model, where the first ranking models of each attribute of time series forecasting models with SVM regression (polynomial kernel; $C=1.0$, $\varepsilon=1.0$) are suitable to the application. The prediction is performed with a trend association-based nearest neighbor predictor. Another article [7] used meteorological data and proposed a long short-term memory-recurrent neural network (LSTM RNN) for dengue disease prediction. Using LSTM has reduced the vanishing gradient problem, which has increased the accuracy. They got 94% accuracy using this model.

The alarming news is that Bangladesh is also at high risk of a dengue fever outbreak. But there are not many resources from Bangladesh available to us for further study [8]. It is high time to take initiative like other countries in Asia to stop this epidemic. This study is designed to work on dengue outbreaks from data collection/preparation to model generation for future forecasting. As such, the authority will have enough time to handle the situation. Many recent works on biomedical

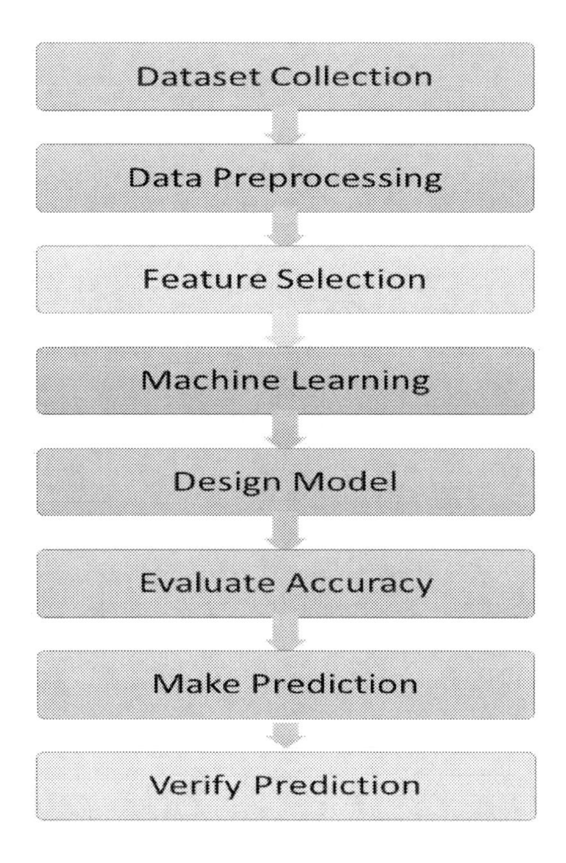

Figure 9.1 The overall workflow diagram to predict dengue incidents.

research have used different types of ensemble techniques in their machine learning methodologies to boost their models' performance [9, 10, 11]. That is why we attempt to employ a weighted average ensemble to further improve our models' predictive ability.

9.3 METHODOLOGY

In our methodology, there are two main building blocks to get expected results as follows.

1. Building a good data preparation tool with feature selection and reduction ability that eventually helps build a prediction model with highest accuracy.

2. Deriving an efficient machine learning model to accurately predict dengue cases to prevent future outbreaks

Our architectural view of the methodology is illustrated in Fig. 9.1 while the tasks in each stage are discussed one by one as follows.

9.3.1 Dataset Collection

We work on meteorological data like temperature, rainfall, humidity, wind speed etc., and epidemiological data like dengue cases. We attempt to build a dengue prediction dataset that contains the number of dengue cases for each month from January 2000 to December 2013 in Dhaka, Bangladesh. Data are collected from the Bangladesh Meteorological Department (website: http://www.bmddataportal.com/) and from the Institute of Epidemiology, Disease Control and Research (IEDCR) (website: https://iedcr.gov.bd/).

There are ten features in this dataset and they are year, month, maximum temperature, minimum temperature, rainfall, relative humidity, wind speed, cloud coverage, bright sunshine, and the number of dengue cases in that month of that year. In total, there are 168 samples in our dataset. We used 80% of the data for training and the remaining 20% are used for testing purposes. Since the size of the dataset is relatively small compared to usual machine learning-based datasets, we have applied k-fold cross-validation on the training dataset to get the best of the data. We put aside the testing dataset for independent testing purposes only. To the best of our knowledge, this is the first dataset that contains both meteorological information and monthly dengue cases in Bangladesh, which makes it a very effective resource to be a potential study reference for further research.

9.3.2 Data Preprocessing

After collecting the data, we preprocess it first in order to apply the machine learning predictors on it. Our data preprocessing task consists of three steps:

- **Feature engineering:** As the number of features in our dataset is relatively low, we desire to apply feature engineering to extract more features related to our task. As such, we extract two new features from the existing data, which are related to the dengue cases. They are the average temperature (AT) and saturated vapor pressure (SVP). They are calculated as follows:

$$AT = \frac{maximumtemperature + minimumtemperature}{2} \quad (9.1)$$

$$SVP = 6.11 * 10^{(7.5*AT)/(237.3+AT)} \quad (9.2)$$

- **Feature selection:** After feature engineering, we select the relevant features for our task. We employ Pearson's correlation coefficient (r) for this purpose, which is calculated as follows:

$$r = \frac{\sum_{i=1}^{n}(x_i - \bar{x})(y_i - \bar{y})}{\sqrt{\sum_{i=1}^{n}(x_i - \bar{x})^2}\sqrt{\sum_{i=1}^{n}(y_i - \bar{y})^2}}. \quad (9.3)$$

  This correlation between features is displayed on a heatmap in Fig. 9.2. According to the heatmap plot, it can be seen that relative humidity is highly correlated with the number of dengue cases. To further demonstrate relative humidity's relation to the number of dengue cases, we provided a couple of

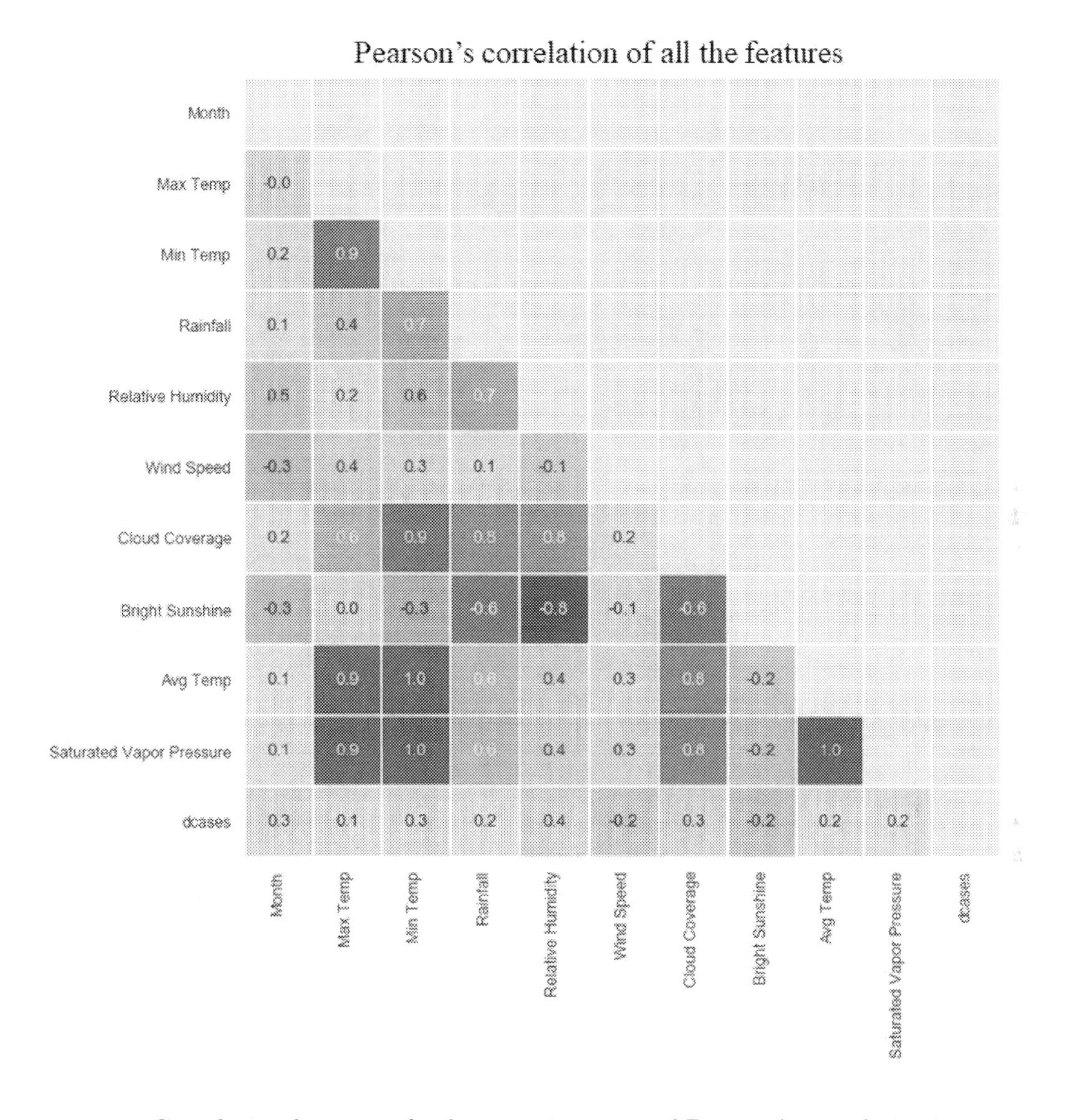

Figure 9.2 Correlation between the features in terms of Pearson's correlation.

graphical illustrations in Fig. 9.3. In Fig. 9.3(a), we show a boxplot to explain this relationship. From the boxplot, we can see which values of relative humidity result in a higher number of dengue cases. We also provided the distribution of relative humidity in our dataset in terms of the number of occurrences in Fig. 9.3(b).

Low correlation often results in bad predictive performance in machine learning methodologies because the lower the correlation the less useful the feature is to the predictor. That is why we removed the three least correlated features, which are maximum temperature, wind speed and brightness of sunshine for further steps. Finally, we choose seven features from all features and apply them to our machine learning models.

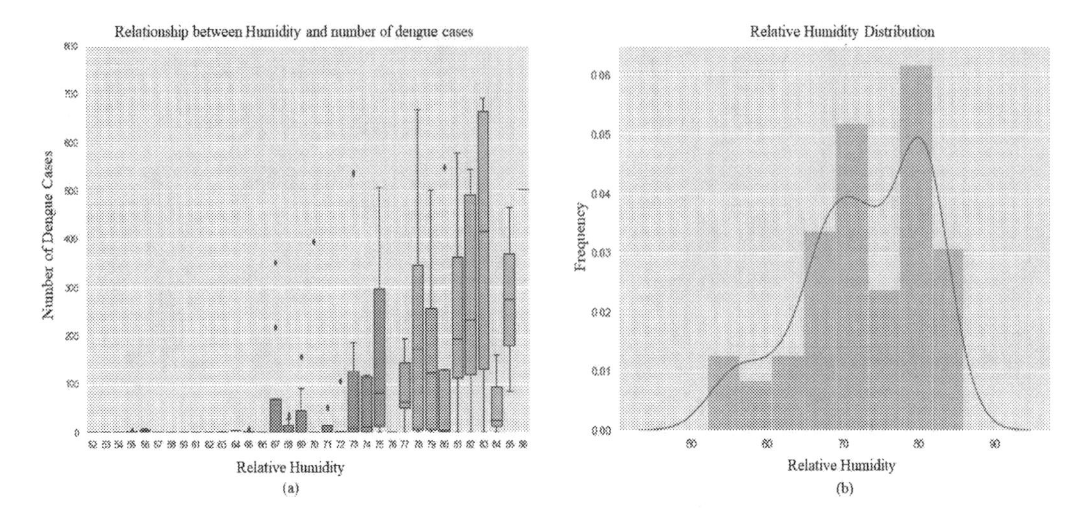

Figure 9.3 (a) Illustration of the relationship between humidity and dengue cases via boxplot. (b) Overall distribution of relative humidity in our dataset.

- **log transformation:** It is very common for the target variable to be right-skewed in regression problems. One of the most popular ways to transform the right-skewed distribution to a "bell-curve" distribution is to apply log transformation to the target variable, in our case the number of dengue cases per month. We can see the distribution of the number of dengue cases before and after the log transformations in Fig. 9.4. In the upper left corner, we can see the right-skewed distribution of the number of dengue cases. This often causes difficulties in learning mechanisms (parameter estimation, convergence time, etc.) in machine learning methodologies. That is why we have applied log transformation in order to achieve a more balanced distribution (upper right corner in Fig. 9.4).

9.3.3 Machine Learning Algorithms

There are some well-established machine learning models for solving regression problems. We decide to take advantage of multiple models in our methodology. That is why we employ the weighted average ensemble technique for the final output calculation. We consider five regression models for our task. They are gradient boosting regression (GBR), extreme gradient boosting regression (XGBR), light gradient boosting regression (LGBR), support vector regressor (SVR), and elastic net. A brief discussion of the used models is given below.

- **Gradient boosting regression (GBR):** GBR has been widely used in various biomedical research because of its ability to combine multiple weak learners into a strong learner by optimizing the loss function iteratively [12, 13]. The parameters involved in building a GBR model: learning rate, maximum depth, number of features at the best split, minimum number of samples at the leaf node, number of estimators, etc.

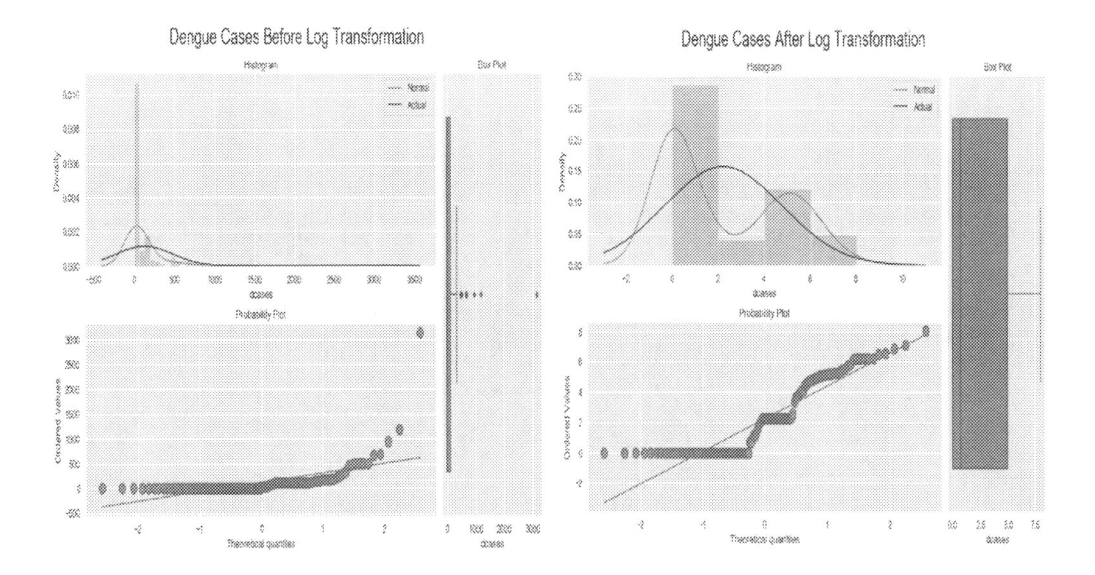

Figure 9.4 Distribution of the target feature (number of dengue cases) before and after log transformation.

- **Extreme gradient boosting regression (XGBR):** XGBR is a popular algorithm in machine learning based studies [14, 15]. In order to overcome the overfitting of GBR, XGBR employs an additive phase in the learning process [16]. It also improves the overall loss function optimization by considering a second-order derivative. It involves the same parameters as the GBR with an extra parameter that defines the minimum loss for the partition of the leaf node.

- **Light gradient boosting regression (LGBR):** LGBR is also a gradient boosting-based algorithm with two significant improvements: removing smaller gradients in estimation and reducing the number of features by wrapping similar features together [17]. Since it is modified from GBR, similar parameters are involved in it as well.

- **Support vector regression (SVR):** SVR is one of the most used regression models in a wide variety of applications because of its ability to select an error margin [18, 19]. We employed the RBF kernel in this task. We needed to define the following parameters to build our SVR model: gamma (kernel width), epsilon and c (regulation parameter).

- **Elastic net regression:** Elastic net is an extension of the linear regression model that combines both $L1$ and $L2$ regularizations to fit the data. It can be used both as an independent model and part of multiple models for ensemble purposes [20, 21, 22]. We tuned two parameters to build this model. They are the $l1$ ratio and alpha (constant to multiply the losses).

In this work, we train and test our data for each of these models independently and apply the weighted average ensemble approach on the models' outputs for the final prediction.

9.3.4 Method Evaluation Metrics

We chose the root-mean-square error (RMSE) to evaluate the performance of our models. RMSE is the squared root of the difference between the actual and predicted value. It is very effective in evaluating the concentration of the output around the actual values. The RMSE is calculated by using the following equation:

$$RMSE = \sqrt{\frac{1}{n}\Sigma_{i=1}^{n}\Big(a_i - p_i\Big)^2} \tag{9.4}$$

Here, $n=$ total number of data points, $a_i=$ actual value, $p_i=$ predicted value.

9.4 RESULT AND DISCUSSION

9.4.1 Parameter Tuning

Parameter optimization is an integral part of building a well-performed methodology. We consider tuning those parameters specifically that can affect the model performance the most. We train our models by applying 10-fold cross-validation using the selected features of the training data. Therefore, nine folds are used for training purposes, while one fold is used for validation. All ten folds are stratified so the distribution of the target class in each fold is equal. The models and training pipeline are implemented using the popular Python library scikit-learn [23]. All training and testing procedures are implemented in a CORE $i5$ laptop with 8GB RAM and 4GB NVIDIA 1050 GPU. The cross-validation and training process take approximately 5–10 minutes to complete for all models. The pipeline is set up in such a way that the best parameters selection and model training are accomplished simultaneously. The selected values after tuning for each model are shown in Table 9.1.

9.4.2 Result Analysis

After tuning and training our models, we test their performance on the testing dataset. Since we apply log transformation as part of the data preprocessing, we need to apply inverse log transformation on the predicted output. We test individually for each model first to get an idea of which model is providing the best performance. This information later helps us in assigning weights during the final prediction. The training and testing RMSE results are provided in Table 9.2. From Table 9.2, we can see that our gradient boosting based model performs relatively better than the SVR and elastic net models. Although our XGBR model provides the lowest RMSE in training, the LGBR model provides the best result during testing. We have also provided the runtime of each algorithm in Table 9.2. We can see the GBR has taken the most and the LGBR model has taken the least time in training.

Table 9.1 Selected values of different parameters of each model after cross-validation

Models	Selected values of parameters
GBR	learning rate=0.01, max_depth=4, min_samples_leaf=5, n_estimators=10
XGBR	learning rate=0.01, max_depth=4, min_child_weight=2, n_estimators=150, gamma=1
LGBR	learning rate=0.001, max_bin=32, num_leaves=5, n_estimators=100
SVR	C=21, epsilon=0.001, gamma=0.01
Elastic net	alpha= 0.0002, l1_ratio=0.8

Table 9.2 Performance of our model on the training and testing datasets

Models	Training RMSE	Testing RMSE	Runtime (seconds)
GBR	0.0894	0.2362	12.23
XGBR	0.0388	0.1084	3.55
LGBR	0.0591	0.0687	1.55
SVR	0.4239	0.6235	0.01
Elastic net	0.5639	0.6558	0.20

Now, in order to calculate the final prediction, we employ the weighted average ensemble technique. The very first step for this is to assign appropriate weights to each of the predictors' outputs according to their performance on the testing dataset. Our approach is to increase weights for the predictors that provided a better result testing dataset and decrease weights for those predictors that performed the opposite, rather than applying the same weight for each predictor. The weights are distributed between 0.10 and 0.30 with 0.05 increments for each model. By applying this, we generate the final predictions using the following equation:

$$final\_output = ((p1 * 0.30)+(p2 * 0.25)+ (p3 * 0.20)+(p4 * 0.15)+(p5 * 0.10))/5 \quad (9.5)$$

where $p1$ = output of the LGBR model, $p2$ = output of the XGBR model, $p3$ = output of the GBR model, $p4$ = output of the SVR model, and $p5$ = output of the elastic net model.

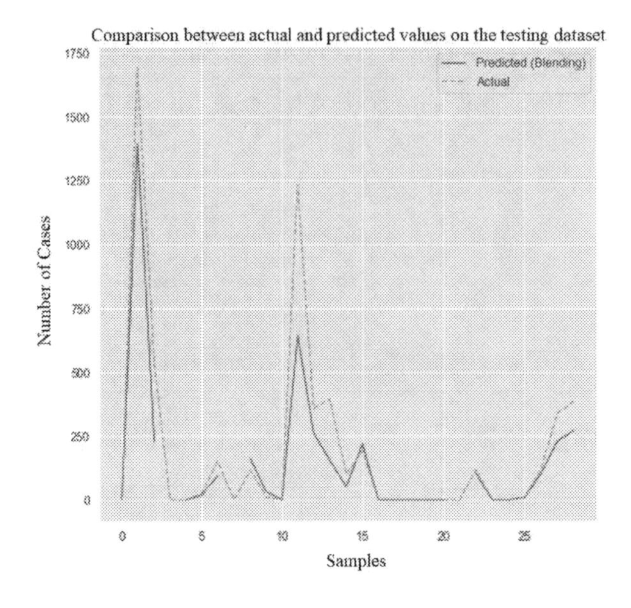

Figure 9.5 Comparison between the predicted and actual result on the test dataset.

Furthermore, in order to demonstrate the performance of our model visually, we provide a comparison between our prediction and the actual output of the test data in Fig. 9.5. From this figure, we can see that our proposed approach is able to predict the spikes in the number of dengue cases quite successfully. Although there are some differences in total numbers, in terms of predicting the upward and downward trends from meteorological data, our model is pretty effective. This prediction can serve as a very valuable tool for taking necessary measures against this viral disease.

Since the dataset we used in this work is curated by us, to the best of our knowledge, this is the first work on dengue prediction from meteorological data in Bangladesh, and that is why we could not find any existing methodology to compare our model's performance. Hence, we believe that our methodology can work as a baseline in future dengue prediction-based works in Bangladesh. Even though we tried our best to provide a fined-tuned methodology, there is still plenty of room for future works. We can explore other machine learning and deep learning algorithms and see their performance. We can also apply our methodology to similar meteorological data-centric prediction tasks and investigate its performance. Moreover, this work not only provides a well-performing predictor for dengue incidence, but also provides vital resources that can be used in further research.

9.5 CONCLUSION

In this work, we have attempted to predict dengue incidence using five state-of-the-art machine learning methodologies. The presented work has the potential to have a huge effect on our medical system. If we get a good prediction of the dengue cases of the upcoming dengue season, it could be helpful for us to take the precautionary measurements. We can redesign our system or upgrade our medical system based on the prediction. If we are able to take precautionary measures based on the prediction, we will hopefully recover from this epidemic situation. Epidemic situations thwart all development works of a nation. Using this prediction module, we can continue all development work of the nation.

ACKNOWLEDGEMENT

The Bangladesh University Grants Commission funded this Project granted to assist the research work of Teachers in Rajshahi University of Engineering & Technology (RUET), under Research Grant no. "DRE/7/RUET/489(31)/ Pro/ 2020-21/10, Dated: 27/07/2020". The authors, therefore, acknowledge with thanks to the Director (Research and Extension) of RUET for technical and financial supports.

Bibliography

[1] Siriyasatien, P., Chadsuthi, S., Jampachaisri, K., and Kesorn, K., 2018. Dengue epidemics prediction: A survey of the state-of-the-art based on data science processes, *IEEE Access*, 6, 53757–53795.

[2] Davi, C., Pastor, A., Oliveira, T., de Lima Neto, F.B., Braga-Neto, U., Bigham, A.W., Bamshad, M., Marques, E.T., and Acioli-Santos, B., 2019. Severe dengue prognosis using human genome data and machine learning, *IEEE Transactions on Biomedical Engineering*, 66 (10), 2861–2868.

[3] Appice, A., Gel, Y.R., Iliev, I., Lyubchich, V., and Malerba, D., 2020. A multi-stage machine learning approach to predict dengue incidence: A case study in Mexico, *IEEE Access*, 8, 52713–52725.

[4] Pham, D.N., Aziz, T., Kohan, A., Nellis, S., Khoo, J.J., Lukose, D., AbuBakar, S., Sattar, A., Ong, H.H., et al., 2018. How to efficiently predict dengue incidence in Kuala Lumpur, in: *2018 Fourth International Conference on Advances in Computing, Communication & Automation* (ICACCA), IEEE, 1–6.

[5] Chellappan, K. et al., 2016. A preliminary dengue fever prediction model based on vital signs and blood profile, in: *2016 IEEE EMBS Conference on Biomedical Engineering and Sciences* (IECBES), IEEE, 652–656.

[6] Nakvisut, A. and Phienthrakul, T., 2018. Two-step prediction technique for dengue outbreak in Thailand, in: *2018 International Electrical Engineering Congress* (iEECON), IEEE, 1–4.

[7] Chovatiya, M., Dhameliya, A., Deokar, J., Gonsalves, J., and Mathur, A., 2019. Prediction of dengue using recurrent neural network, in: *2019 3rd International Conference on Trends in Electronics and Informatics* (ICOEI), IEEE, 926–929.

[8] Dhaka, A. and Singh, P., 2020. Comparative analysis of epidemic alert system using machine learning for dengue and chikungunya, in: *2020 10th International Conference on Cloud Computing, Data Science & Engineering* (Confluence), IEEE, 798–804.

[9] Kaur, H., Malhi, A.K., and Pannu, H.S., 2020. Machine learning ensemble for neurological disorders, *Neural Computing and Applications*, 1–18.

[10] Shakeel, P.M., Tolba, A., Al-Makhadmeh, Z., and Jaber, M.M., 2020. Automatic detection of lung cancer from biomedical data set using discrete AdaBoost optimized ensemble learning generalized neural networks, *Neural Computing and Applications*, 32 (3), 777–790.

[11] Abba, S.I., Linh, N.T.T., Abdullahi, J., Ali, S.I.A., Pham, Q.B., Abdulkadir, R.A., Costache, R., Anh, D.T., et al., 2020. Hybrid machine learning ensemble techniques for modeling dissolved oxygen concentration, *IEEE Access*, 8, 157218–157237.

[12] Li, X., Li, W., and Xu, Y., 2018. Human age prediction based on DNA methylation using a gradient boosting regressor, *Genes*, 9 (9).

[13] Wang, H. and Gu, G., 2015. Wavelet gradient boosting regression method study in short-term load forecasting, *Smart Grid*, 5 (4), 189–196.

[14] Bagalkot, N., Keprate, A., and Orderl∅kken, R., 2021. Combining computational fluid dynamics and gradient boosting regressor for predicting force distribution on horizontal axis wind turbine, *Vibration*, 4 (1), 248–262.

[15] Acharya, S., Swaminathan, D., Das, S., Kansara, K., Chakraborty, S., Kumar, D., Francis, T., and Aatre, K.R., 2020. Erratum: Non-invasive estimation of hemoglobin using a multimodel stacking regressor, *IEEE Journal of Biomedical and Health Informatics*, 24 (9), 2726.

[16] Chen, T. and Guestrin, C., 2016. Xgboost: A scalable tree boosting system, in: *Proceedings of the 22nd acm sigkdd international conference on knowledge discovery and data mining*, 785–794.

[17] Ke, G., Meng, Q., Finley, T., Wang, T., Chen, W., Ma, W., Ye, Q., and Liu, T.Y., 2017. Lightgbm: A highly efficient gradient boosting decision tree, *Advances in neural information processing systems*, 30, 3146–3154.

[18] Das, H., Naik, B., and Behera, H., 2020. An experimental analysis of machine learning classification algorithms on biomedical data, in: *Proceedings of the 2nd International Conference on Communication, Devices and Computing*, Springer, 525–539.

[19] Deng, F., Huang, J., Yuan, X., Cheng, C., and Zhang, L., 2021. Performance and efficiency of machine learning algorithms for analyzing rectangular biomedical data, *Laboratory Investigation*, 101 (4), 430–441.

[20] Rauschenberger, A., Glaab, E., and van de Wiel, M., 2020. Predictive and interpretable models via the stacked elastic net, *Bioinformatics*.

[21] Marafino, B.J., Boscardin, W.J., and Dudley, R.A., 2015. Efficient and sparse feature selection for biomedical text classification via the elastic net: Application to ICU risk stratification from nursing notes, *Journal of Biomedical Informatics*, 54, 114–120.

[22] Acharya, S., Swaminathan, D., Das, S., Kansara, K., Chakraborty, S., Kumar, D., Francis, T., and Aatre, K.R., 2020. Erratum: Non-invasive estimation of hemoglobin using a multimodel stacking regressor, IEEE Journal of Biomedical and Health Informatics, 24 (9), 2726.

[23] Pedregosa, F., Varoquaux, G., Gramfort, A., Michel, V., Thirion, B., Grisel, O., Blondel, M., Prettenhofer, P., Weiss, R., Dubourg, V., et al., 2011. Scikit-learn: Machine learning in Python, the *Journal of Machine Learning Research*, 12, 2825–2830.

CHAPTER 10

Detecting DNS over HTTPS Traffic Using Ensemble Feature-based Machine Learning

Sajal Saha

Patuakhali Science and Technology University, Patuakhali, Bangladesh

Moinul Islam Sayed

Patuakhali Science and Technology University, Patuakhali, Bangladesh

Rejwana Islam

Global University Bangladesh, Barishal, Bangladesh

CONTENTS

CYBERCRIMINALS use various tools, including malware, phishing, ransomware, and denial of service, among others, to conduct cyberattacks. Attackers also use the Domain Name System (DNS) as an unpredictable cybercrime medium. To prevent this kind of attack, Hypertext Transfer Protocol Secure (HTTPS) plays a significant role that secures computer communication over the network. An alternative solution has been proposed by security organizations where the DNS requests travel over HTTPS, and the concept is known as DNS over HTTPS (DoH). However, the attackers use this new protocol to inject their data in an encrypted way the is undetectable by firewalls and other methods. The Canadian Institute of Cyber Security recently published a dataset, CIRA-CIC-DoHBrw-2020 [8], consisting of malicious and benign DoH traffic. This chapter proposes a machine learning solution based on an ensemble feature selection technique to classify the DoH traffic. We experimented with several machine learning models feeding different feature sets extracted from different feature selection algorithms. According to our analysis, the ensemble feature-based machine learning model outperforms the other models based on the individual feature set.

10.1 INTRODUCTION

The Domain Name System (DNS) provides a mapping service from a human-understandable domain name to the machine-readable Internet Protocol (IP) address. DNS helps to load the internet resources by converting the domain name to a specific machine address, which is unique for every device connected to the internet. There are four different DNS servers to query the requested host IP address, and they are called only if the requested query data is not available in the cache memory. The recursive DNS resolver repetitively searches for the host IP address in DNS recursor, Root nameserver, Top Level Domain (TLD) nameserver, and Authoritative nameserver until it gets a confident answer.

A cyberattack exploits DNS requests and responses to transfer the encoded information of other processes and protocols. The procedure is called DNS tunneling and facilitates the attacker's use of DNS traffic for cyberattacks. Typically, most firewalls do not analyze the DNS packet data to identify abnormal activities, which is one of the most important reasons to use DNS as a method of data injection. Also, there are different standard software tools available in the market that help generate a DNS tunneling attack without in-depth technical knowledge, which is another reason for choosing DNS queries to mount cyberattacks.

Recently, a new protocol has been invented and proposed as an internet standard (IETF RFC8484) which processes the DNS request using secure communication [3]. This protocol is called DNS-over-HTTPS and is illustrated in Figure 10.1, and it encrypts the request and response data securely. Generally, the DNS request is placed in plaintext, which is visible in the network, and there is a chance of a man-in-the-middle attack. However, due to the enhanced secured communication mechanism of DoH, the possibility of such attacks has been reduced.

Although the encryption mechanism of the DoH protocol helps diminish some attacks, it creates few other challenges that assist attackers in different ways. For example, if a client tries to access a lousy domain, it can be easily identified by analyzing DNS request data. But this new encrypted protocol impedes this kind of analysis to classify the abnormal activity. This protocol impacts the enterprise sector because the system administrator generally uses the DNS-based solution to

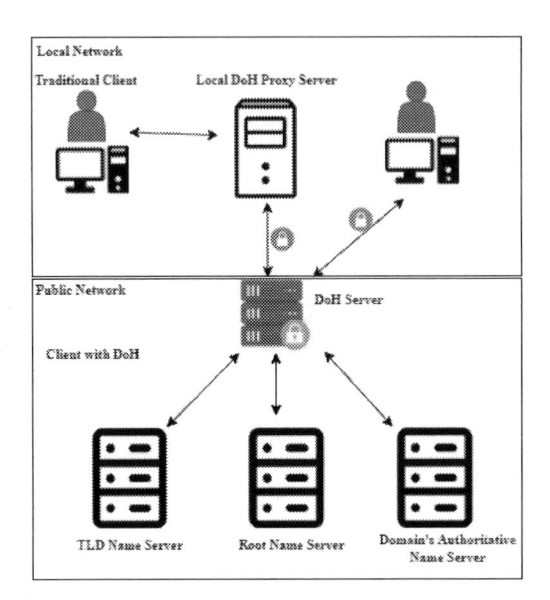

Figure 10.1 DNS over HTTPS (DoH).

filter and monitor traffic to avoid workers using unrelated sites and malware hosts. An attacker can take advantage of the encrypted communication of DoH to send their data, which will be unidentified by the network. Also, the DoH protocol enables attackers to send several requests without initiating several connections, which helps to fleece the frequency of their DNS resolution.

In this research, we propose a machine learning solution to identify the bad DoH request. An ensemble feature selection technique has selected an optimal feature set for our Machine Learning (ML) models. A total of seven different Feature Selection (FS) methods were implemented to choose the individual feature set from the dataset, which was then used to classify benign and malignant DoH requests. The individual feature subsets were ensembled using the Majority Voting (MV) technique for choosing the best feature subset. Finally, we compare the performance of our model for individual feature sets and the ensemble feature set to determine the best model.

This chapter is organized as follows. Section 10.2 describes the literature review of current DoH traffic systems based on machine learning models. Section 10.3 presents the proposed methodology to design the ensemble feature selection-based DoH traffic detection. Section 10.4 presents the performance results of the different FS methods in ML models. Section 10.5 concludes our chapter and outlines the ideas for future research actions.

10.2 LITERATURE REVIEW

Reference [2] discussed DoH in general including privacy, security, and performance of the architecture. Reference [6] offered a concise overview of DoH operation and define DoH behavioral trends using the RITA (Real Intelligence Threat Analytics) framework. Reference [9] raised the question about the privacy of DoH, claiming that, based on the pattern of the DoH packet size, detection of DoH resolvers is possible.

Reference [14] examined DoH and DoT traffic to develop an innovative approach to website fingerprinting. It proposed a model that uses packet sizes as an important feature for classifying encrypted DoH traffic. They created fingerprints for every website using the patterns of encrypted DNS traffic. Reference [4] conducted a study to recognize DoH content in encrypted traffic. They used some key feature vectors to evaluate the classifier's performance. To analyze DoT traffic produced by visiting websites, Reference [7] developed a DoT fingerprinting technique that decides whether a DNS request is genuine by distinguishing between a real user and an attacker.

Reference [12] used supervised machine learning for analyzing DNS traffic and identifying benign and malicious domains. It extracted features from subdomains in different groups. To classify the DNS traffic, it filtered the primary domain rather than queries. The classification of traffic employed Random Forest, AdaBoost, Gradient Boosting, Bagging, SVC, and SGD. This method, however, has the drawback of being unable to detect malicious queries in the main domain.

Reference [1] used several machine learning algorithms such as (Gradient Boosting, XG Boost, and Light Gradient Boosting Machine) to examine a dataset called CIRA-CIC-DoHBrw-2020, which is also used in our experiment. The author claimed that source IP address and destination IP address, length of the packet, Mode and Median of the packet, and skew from mode are the significant features in the whole dataset. This study concluded that XG Boost and Light Gradient Boosting Machine outperformed the other algorithms in nearly all classification metrics.

Reference [13] considered network traffic over the course of a single day, focusing solely on reflector traffic. They used SVM (Support Vector Machine) to classify benign and malicious DNS servers. However, this method is unable to deal with encrypted communication. [10] aimed at using DGAs (Domain Generation Algorithms) to mask malicious DNS queries. The Hodrick–Prescott filter was used to identify traffic created by several DGAs.

Reference [16] used machine learning to distinguish between DoH and nonDoH requests. They used five machine learning classifiers in the experiment and were able to detect DoH traffic with 99.9% accuracy. [15] used generated DoH malicious and benevolent traffic datasets to detect DoH attacks. The dataset contained 269,299 samples and the author claimed that the best training-to-test ratio was 1:3. Among the five classifiers: Naive Bayes, Logistic regression, Random Forest, K Nearest Neighbour, Gradient Boosting, Random Forest and Gradient Boosting both show the maximum accuracy.

10.3 METHODOLOGY

10.3.1 Dataset

We used the CIRA-CIC-DoHBrw-2020 dataset [8] created by the Canadian Institute for Cybersecurity. They generated data in two different layers by accessing a considerable number (10k) of Alexa websites. DNS tunneling tools have been used to produce malicious requests. The traffic is classified as DoH and non-DoH at the first

Table 10.1 Dataset summary.

	Total instance	Malicious	Benign
Train dataset	26600	13300	13300
Test dataset	11400	5700	5700

layer, while the DoH traffic is later classified as malicious and benign at layer 2. We used the layer 2 dataset for our analysis, and there are two different datasets named l2-benign.csv and l2-malicious.csv, which contain benign and malicious DoH traffic, respectively. There is a total of 249836 instances of malicious traffic in the data set, while the total number of benign traffic instances is 19807. We collected a total of 38000 traffic instances for our analysis, equally divided into benign and malicious traffic. Since we applied supervised learning, the equal distribution of label data is considered for our experiment, although it is not mandatory. The selected dataset is then divided into 70% / 30% as training and testing data, respectively. So the total number of training instances is 26600, and testing records is 11400. The summary of the dataset considered for the experiment has been shown in Table 10.1.

10.3.2 Data Preprocessing

Data preprocessing is the very first and crucial step of the machine learning model. The model performance is directly related to the quality of the data [5]. So we need to preprocess our data before feeding it into the classification model. Data preprocessing techniques help remove the null and infinity values, convert data into a trainable form, normalize and scale data, etc., which impact the ultimate performance of the model. Our benign and malicious datasets both contain null values. The null values were comparatively less than the total number of instances, so the null values were removed from the dataset for our experiment, although, there are different techniques, e.g., replace null values with the mean or median, to handle null values in a small dataset. It is essential to convert categorical data into numerical values as categorical data is incompatible with a machine learning model. Our experiment used the LabelEncoder from the Scikit Learn library [11] to convert our categorical data into numerical values. There is another popular encoding technique called one-hot encoding to process categorical information. Finally, the data has been normalized so that it cannot be biased by the features with a large value range that might make a wrong inference. We used a technique called Min-Max scaling in our experiment to normalize our data into ranges 0 and 1.

10.3.3 Feature Engineering

In a machine learning model, the data can have a long list of not equally important features to design a best-performing classifier or regressor. The irrelevant features not only extend the training time but also have an effect on the model's efficiency. So it is an essential step to extract the best feature set that can identify the target efficiently and accurately. A total of seven different feature selection algorithms were implemented in our experiment to investigate the CIRA-CIC-DoHBrw-2020 dataset. In our proposed solution, we applied an ensemble-based FS approach explained later to produce a majority-selected feature set based on the contribution of individual feature selection methods. In terms of accuracy, this ensemble FS technique outperforms the individual FS technique.

10.3.4 Machine Learning Models

In this step, individual feature sets collected using seven feature selection methods, a full feature set with any feature engineering, and an ensemble feature set combining the individual feature selection results were used to implement various machine learning models. For example, we used different numbers of best features in our FS methods, such as 10, 15, and 20, and found that the parameter value 20 gave us the best results. Five different machine learning algorithms were implemented in our proposed solution to assess output based on various feature sets. We used Naive Bayes (NB), Logistic Regress (LR), Neural Network (NN), Decision Tree (DT), Random Forest (RF), and Support Vector Machine (SVM).

10.3.5 Proposed DoH Detection Model

Our proposed ensemble feature selection-based DoH flow classification technique has five major units: data preprocessing, feature selection using individual feature selection system, generate ensemble feature collection, training and testing the machine learning model, and performance analysis. To compare their results, we used seven different FS methods to extract the set of features to input into the classifier. To verify the model classification, the same feature set from the test set has been used.

The ensemble feature set was constructed by combining all of the feature sets selected by the individual FS process. This feature set was also used to train and evaluate our classification models, allowing us to compare them to individual FS methods. Finally, we used all of the performance metrics to classify our finest feature set and the ML model that goes with it.

10.3.5.1 Ensemble Feature Selection

Filter-based FS methods, Wrapper-based FS methods, and Embedded FS methods are the three major types of FS methods. We used a total of seven FS algorithms in our experiment, mixing various forms. Pearson's correlation (PEARSON) and ANOVA are two of the seven feature selection methods, which is the filter-based method used in our analysis. As a Wrapper-based approach, Mutual Information (MUTINFO) and Recursive Feature Elimination (RFE) are utilized. Finally, as embedded FS methods, we chose LASSO Regression (LASSO), Logistic Regression (LGR), and Random Forests (RF). We selected features that have been agreed on by more than half of the feature selection techniques (e.g., we considered the features chosen by four or more algorithms out of seven).

10.3.6 Software and Hardware Preliminaries

We used Python, and the machine learning library scikit-learn to conduct the experiments in computers with the following configuration: Intel (R) i3-8130U CPU@2.20GHz, 8GB memory, and 64-bit windows operating system.

10.3.7 Evaluation Metrics

We used five different evaluation metrics, e.g., Accuracy, Precision, Recall, F-1 Score, and False Positive Rate to estimate the performance of our classification models. There are four measurement parameters, True Positive (TP), False Positive (FP), True Negative (TN), and False Negative (FN), which are used to define the evaluation metrics stated above. Besides, we used another evaluation metric called Area Under the Curve–Receiver Operating Characteristics (AUC-ROC) curve to measure our classification model's performance. This curve illustrates the performance at various threshold settings. ROC is a probability curve, and AUC represents the degree or measure of separability based on the False Positive Rate (FPR) and True Positive Rate (TPR). The ROC curve is plotted with TPR in the y-axis vs. FPR in the x-axis and classification; the higher the ROC, the better the model is at distinguishing the classes. By analogy, the higher the AUC, the better the model determines malicious DoH traffic.

10.4 RESULTS AND DISCUSSION

The experimental results are analyzed and described here to evaluate the performance of our proposed framework. Different feature selection methods selected a set of significant features. We applied the ensemble technique (i.e., Majority Voting) to figure out the best set of features chosen by most individual feature selection methods. Then, we identified and compared the best performing machine learning model for each feature set determined by particular feature selection algorithms. The features derived from seven feature selection methods are shown in Table 10.2.

On the feature list in Table 10.2 that was picked using seven feature selection methods, we used the majority voting technique. When more than half of the feature selector candidates choose a feature, it becomes more relevant. Our ensemble

Table 10.2 Extracted features from 7 FS methods.

FS Method	Selected Features
Pearson	['ResponseTimeTimeMean', 'PacketTimeMedian', 'ResponseTimeTimeCoefficientofVariation', 'ResponseTimeTimeMedian', 'PacketTimeStandardDeviation', 'PacketLengthSkewFromMedian', 'PacketTimeMean' , 'ResponseTimeTimeSkewFromMedian', 'FlowBytesSent', 'PacketLengthMode', 'PacketTimeSkewFromMode', 'F lowBytesReceived', 'Duration', 'PacketTimeCoefficientofVariation', 'PacketLengthVariance',

(*Continued on next page*)

Table 10.2 (Continued)

FS Method	Selected Features
	'ResponseTimeTimeSkewFromMode', 'PacketTimeSkewFromMedian', 'PacketLengthMean', 'PacketLengthStandardDeviation', 'PacketLengthCoefficientofVariation']
Mutual Info	['FlowBytesSent', 'FlowBytesReceived', 'PacketLengthVariance', 'Packet LengthStandardDeviation', 'PacketLengthMedian', 'PacketLengthMode']
ANOVA	['Duration', 'FlowBytesSent', 'FlowBytesReceived', 'PacketLengthVariance', 'PacketLengthStandardDeviation', 'PacketLengthMean', 'PacketLengthMode', 'PacketLengthSkewFromMedian', 'PacketLengthCoefficientofVariation', 'PacketTimeStandardDeviation', 'PacketTimeMean', 'PacketTimeMedian', 'PacketTimeSkewFromMedian', 'PacketTimeSkewFromMode', 'PacketTimeCoefficientofVariation', 'ResponseTimeTimeMean', 'ResponseTimeTimeMedian', 'ResponseTimeTimeSkewFromMedian', 'ResponseTimeTimeSkewFromMode', 'ResponseTimeTimeCoefficientofVariation']
LRL1	['Duration', 'FlowBytesSent', 'FlowReceivedRate', 'PacketLengthVariance', 'PacketLengthStandardDeviation', 'PacketLengthMean', 'PacketLengthMedian', 'PacketLengthSkewFromMedian', 'PacketLengthSkewFromMode', 'PacketLengthCoefficientofVariation', 'PacketTimeVariance', 'PacketTimeStandardDeviation', 'PacketTimeMode', 'PacketTimeCoefficientofVariation', 'ResponseTimeTimeVariance', 'ResponseTimeTimeStandardDeviation', 'ResponseTimeTimeMean', 'ResponseTimeTimeMedian', 'ResponseTimeTimeSkewFromMedian', 'ResponseTimeTimeCoefficientofVariation']
LASSO	['Duration', 'FlowBytesSent', 'FlowBytesReceived', 'PacketLengthVariance', 'PacketLengthStandardDeviation', 'PacketLengthMean',

(*Continued on next page*)

Table 10.2 (Continued)

FS Method	Selected Features
	'ResponseTimeTimeMode', 'PacketTimeVariance', 'PacketTimeStandardDeviation', 'ResponseTimeTimeStandardDeviation', 'ResponseTimeTimeMedian', 'PacketLengthCoefficientofVariation']
RF	['Duration', 'FlowBytesSent', 'FlowBytesReceived', 'PacketLengthVariance', 'PacketLengthStandardDeviation', 'PacketLengthMean', 'PacketLengthMedian', 'PacketLengthMode', 'PacketLengthSkewFromMode', 'PacketLengthCoefficientofVariation', 'PacketTimeVariance', 'PacketTimeStandardDeviation']
RFE	['Duration', 'FlowBytesSent', 'FlowBytesReceived', 'FlowReceivedRate', 'PacketLengthVariance', 'PacketLengthStandardDeviation', 'PacketLengthMean', 'PacketLengthSkewFromMode', 'PacketLengthCoefficientofVariation', 'PacketTimeVariance', 'PacketTimeStandardDeviation', 'PacketTimeMean','PacketTimeMedian', 'PacketTimeMode', 'ResponseTimeTimeVariance', 'ResponseTimeTimeStandardDeviation', 'ResponseTimeTimeMean', 'ResponseTimeTimeMedian', 'ResponseTimeTimeSkewFromMedian', 'ResponseTimeTimeCoefficientofVariation']

architecture only considers features selected by at least four feature selectors since we used seven individual feature selectors. Table 10.3 shows that our system selects 13 features. Each feature is denoted by a Yes (selected) or a No (not selected) to indicate whether it was chosen using the individual feature selection process. In addition, the vote column in Table 10.3 represents the total number of FS methods which select this feature as important. The feature ranking based on total number of votes is illustrated in Figure 10.2.

Table 10.3 Extracted features using the ensemble feature selection method.

Feature	PEARSON	MUTINFO	ANOVA	LRL1	LASSO	RF	RFE	Vote
FlowBytesSent	Yes	Yes	Yes	Yes	Yes	Yes	Yes	7
FlowBytesReceived	Yes	Yes	Yes	Yes	Yes	Yes	Yes	7
PacketLengthVariance	Yes	Yes	Yes	Yes	Yes	Yes	Yes	7
PacketLengthStandardDeviation	Yes	Yes	Yes	Yes	Yes	Yes	Yes	7
Duration	Yes	No	Yes	Yes	Yes	Yes	Yes	6
PacketLengthMean	Yes	No	Yes	Yes	Yes	Yes	Yes	6
PacketLengthCoefficientofVariation	Yes	No	Yes	Yes	Yes	Yes	Yes	6
PacketTimeStandardDeviation	Yes	No	Yes	Yes	Yes	No	Yes	5
ResponseTimeTimeMedian	Yes	No	Yes	Yes	Yes	No	Yes	5
PacketLengthMode	Yes	Yes	Yes	No	No	Yes	No	4
PacketLengthSkewFromMedian	Yes	No	Yes	Yes	No	No	Yes	4
PacketTimeVariance	No	No	No	Yes	Yes	Yes	Yes	4
PacketTimeMean	Yes	No	Yes	No	No	Yes	Yes	4
ResponseTimeTimeMean	Yes	No	Yes	Yes	No	No	Yes	4
ResponseTimeTimeSkewFromMedian	Yes	No	Yes	Yes	No	No	Yes	4
ResponseTimeTimeSkewFromMode	Yes	No	Yes	Yes	No	No	Yes	4

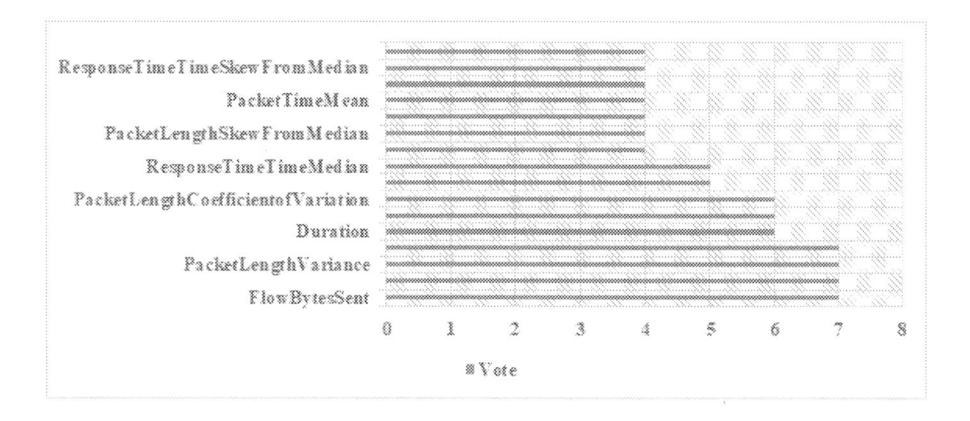

Figure 10.2 Feature ranking.

Five different ML models have been trained on the individual feature set selected by FS methods. The ensemble feature set determined by the MV technique is presented in Table 10.3. The model was also trained using the full feature set without any feature engineering. The best performing model for each feature set was established, and the results are shown in Table 10.4. Based on the ensemble feature set, we found that the DT model performs best, with a highest accuracy of 99% percent and lowest FPR of 0.01. Figure 10.3 depicts a schematic representation of the various

Table 10.4 Best performing classifier with corresponding feature set and performance metrics

Classifier	Feature	Acc.	F1	Prec.	Recall	FPR	TPR	Train Time
NB	ALL	0.84	0.85	0.84	0.84	0.10	0.79	0.84
NN	RF	0.97	0.97	0.97	0.97	0.04	0.98	0.97
SVM	RF	0.95	0.95	0.95	0.95	0.08	0.98	0.95
LR	LASSO	0.89	0.89	0.89	0.89	0.15	0.92	0.89
DT	EN	0.99	0.99	0.99	0.99	0.01	1.00	0.99

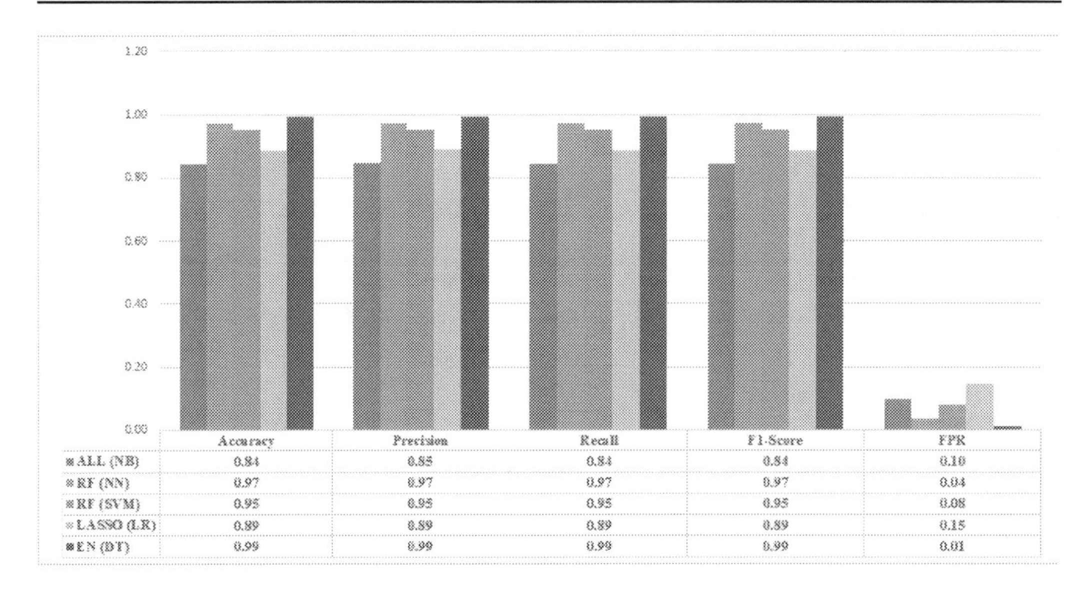

Figure 10.3 Performance analysis.

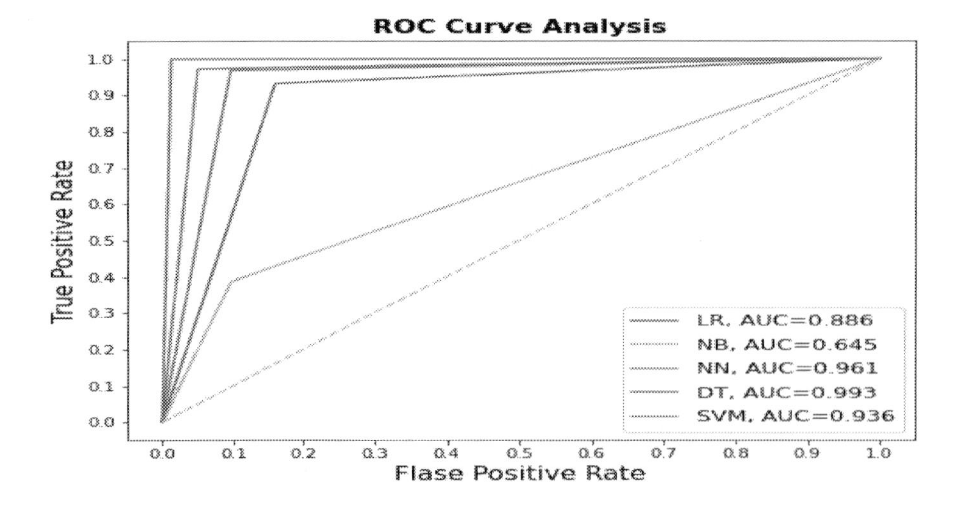

Figure 10.4 ROC curve.

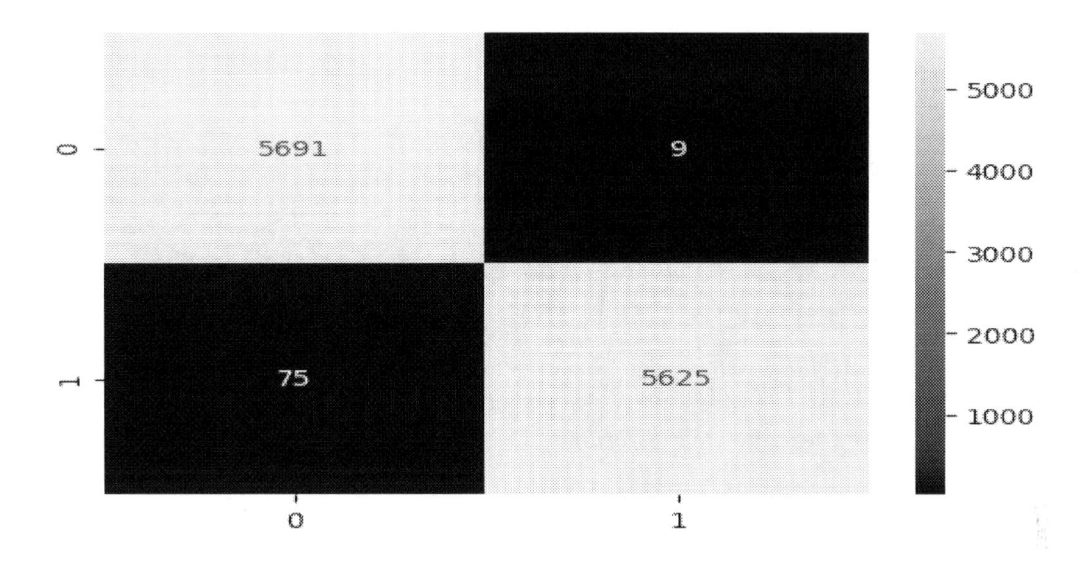

Figure 10.5 Confusion matrix for best performing model.

performance metrics based on their feature sets. The receiver operating curve for all machine learning models based on the ensemble feature set is shown in Figure 10.4 because it outperforms the other feature selection process. Figure 10.5 represents the confusion matrix for our best performing model which is the decision tree with ensemble feature set in Table 10.4.

10.5 CONCLUSION

Our findings demonstrate the importance of choosing the best feature set while using machine learning techniques to detect DoH requests. This research presented a comparative analysis among the feature sets selected by seven different FS methods, ensemble feature sets, and complete feature sets. Our designed machine learning model found the ensemble feature set more suitable to classify the malicious and benign DoH traffic. We created 5 different machine learning models based on a total of 9 feature sets. In this way, we trained a total of 45 classification models. After measuring their performance, we concluded that the decision tree classifier with the ensemble feature set performs better than the other models in terms of accuracy and FPR. As future work, we would like to apply this technique using deep neural networks to do comparative work between machine learning and deep learning models.

Bibliography

[1] Banadaki, Y. M. (2020). Detecting Malicious DNS over HTTPS Traffic in Domain Name System using Machine Learning Classifiers. *Journal of Computer Sciences and Applications*, 8(2), 46–55.

[2] Borgolte, K., Chattopadhyay, T., Feamster, N., Kshirsagar, M., Holland, J., Hounsel, A., & Schmitt, P. (2019). How DNS over HTTPS is reshaping privacy,

performance, and policy in the internet ecosystem. *Performance, and Policy in the Internet Ecosystem* (July 27, 2019).

[3] Böttger, T., Cuadrado, F., Antichi, G., Fernandes, E. L., Tyson, G., Castro, I., & Uhlig, S. (2019, October). An Empirical Study of the Cost of DNS-over-HTTPS. *In Proceedings of the Internet Measurement Conference* (pp. 15–21).

[4] Bushart, J., & Rossow, C. (2020). Padding Ain't Enough: Assessing the Privacy Guarantees of Encrypted DNS. *In 10th USENIX Workshop on Free and Open Communications on the Internet (FOCI 20).*

[5] Kotsiantis, S. B., Kanellopoulos, D., & Pintelas, P. E. (2006). Data preprocessing for supervised learning. *International Journal of Computer Science*, 1(2), 111–117.

[6] Hjelm, D. (2019). A new needle and haystack: Detecting DNS over HTTPS usage. *SANS Institute, Information Security Reading Room*, August.

[7] Houser, R., Li, Z., Cotton, C., & Wang, H. (2019, December). An investigation on information leakage of DNS over TLS. *In Proceedings of the 15th International Conference on Emerging Networking Experiments And Technologies* (pp. 123–137).

[8] MontazeriShatoori, M., Davidson, L., Kaur, G., and Lashkari, A. H. (2020, August). Detection of doh tunnels using time-series classification of encrypted traffic. *In 2020 IEEE Intl Conf on Dependable, Autonomic and Secure Computing, Intl Conf on Pervasive Intelligence and Computing, Intl Conf on Cloud and Big Data Computing, Intl Conf on Cyber Science and Technology Congress (DASC/PiCom/CBDCom/CyberSciTech)* (pp. 63–70). IEEE.

[9] Nijeboer, F. J. (2020). Detection of HTTPS encrypted DNS traffic (Bachelor's thesis, University of Twente).

[10] Patsakis, C., Casino, F., & Katos, V. (2020). Encrypted and covert DNS queries for botnets: Challenges and countermeasures. *Computers & Security*, 88, 101614.

[11] Pedregosa, F., Varoquaux, G., Gramfort, A., Michel, V., Thirion, B., Grisel, O., ... & Duchesnay, E. (2011). Scikit-learn: Machine learning in Python. *Journal of Machine Learning Research*, 12, 2825–2830.

[12] Preston, R. (2019, November). DNS Tunneling Detection with Supervised Learning. *In 2019 IEEE International Symposium on Technologies for Homeland Security (HST)* (pp. 1–6). IEEE.

[13] Shima, K., Nakamura, R., Okada, K., Ishihara, T., Miyamoto, D., & Sekiya, Y. (2019, December). Classifying DNS Servers Based on Response Message Matrix Using Machine Learning. *In 2019 International Conference on Computational Science and Computational Intelligence (CSCI)* (pp. 1550–1551). IEEE.

[14] Siby, S., Juarez, M., Diaz, C., Vallina-Rodriguez, N., & Troncoso, C. (2019). Encrypted DNS --> Privacy? A Traffic Analysis Perspective. arXiv preprint arXiv:1906.09682.

[15] Singh, S. K., & Roy, P. K. (2020, December). Detecting Malicious DNS over HTTPS Traffic Using Machine Learning. *In 2020 International Conference on Innovation and Intelligence for Informatics, Computing and Technologies (3ICT)* (pp. 1–6). IEEE.

[16] Vekshin, D., Hynek, K., & Cejka, T. (2020, August). Doh insight: Detecting DNS over HTTPS by machine learning. *In Proceedings of the 15th International Conference on Availability, Reliability and Security* (pp. 1–8).

CHAPTER 11

Development of a Risk-Free COVID-19 Screening Algorithm from Routine Blood Tests Using Ensemble Machine Learning

Md. Mohsin Sarker Raihan

Dept. of BME, Khulna University of Engineering and Technology, Khulna, Bangladesh

Md. Mohi Uddin Khan

Dept. of EEE, Islamic University of Technology, Boardbazar, Gazipur, Bangladesh

Laboni Akter

Dept. of BME, Khulna University of Engineering and Technology, Khulna, Bangladesh

Abdullah Bin Shams

Dept. of ECE, University of Toronto, Toronto, Ontario, Canada

CONTENTS

THE REVERSE TRANSCRIPTION POLYMERASE CHAIN REACTION (RTPCR) test is the silver bullet diagnostic test to discern COVID infection. Rapid antigen detection is a screening test to identify COVID positive patients in as little as 15 minutes, but has a lower sensitivity than the PCR tests. Besides having multiple standardized test kits, many people are getting infected and either recovering or dying even before the test due to the shortage and cost of kits, lack of indispensable specialists and labs, and time-consuming results compared to the bulk population especially in developing and underdeveloped countries. Intrigued by the parametric deviations in immunological and hematological profiles of a COVID patient, this research work leveraged the concept of COVID-19 detection by proposing a risk-free and highly accurate Stacked Ensemble Machine Learning model to identify a COVID patient from communally available, widespread, and cheap routine blood tests that give promising accuracy, precision, recall and F1-score of 100%. Analysis from R-curves also show the preciseness of the risk-free model to be implemented. The proposed method has the potential as a large-scale ubiquitous low-cost screening application. This can add an extra layer of protection in keeping the number of infected cases to a minimum and controlling the pandemic by identifying asymptomatic or pre-symptomatic people early.

11.1 INTRODUCTION

The pandemic by SARS-CoV-2 infection has claimed over 4 million lives around the world to this date. Since the outbreak, countries have expeditiously ramped up their capacity in full for testing and contact tracing, and have declared a 195.545 million confirmed cases from 2.35 billion tests conducted worldwide until now [1]. The sub-atomic experiment accomplished using the converse polymerase chain response (PCR) method is the tool of choice, or the best quality standard, for detecting SARS-CoV-2 contamination. However, the test is time-consuming, necessitating the use of specialized hardware and substances, collaboration of particular and qualified recruits designed for the sample assortment, and relying on the sufficient hereditary security of the RNA groupings chosen for tempering the preliminary [2]. Attempts to develop tests based on IgM/IgG antibodies also suffer from low sensitivity, specificity and costly reagents.

To improve symptomatic capacities, the information science local area has suggested a few AI machine learning (ML) models. The majority of these models depend on processed tomography outputs or chest X-rays. Despite the detailed promising outcomes, there are some significant concerns in regard to these and different works, particularly as to arrangements dependent on chest X-rays, which have been related to a high number of negative outcomes. Then again, arrangements dependent on CT imaging are influenced by the attributes of this methodology: CTs are exorbitant, tedious, and need particular devices; accordingly, methodologies dependent on this imaging procedure cannot be applied sensibly for screening tests. Detection based on chest CT images and X-rays cannot be performed abundantly due to even higher cost and radiation exposure.

Considering the detrimental effects of nCoV breeding inside host lung cells utilizing the Angiotensin-Converting Enzyme 2 (ACE2) found in ample amounts on type-II alveolar cells and considering inflammation in the lungs and respiratory tracts, recent studies have reported a significant change in immunological and hematological parameters in the host blood stream. The predominant purpose of this study is to develop a novel machine learning technique, featuring diagnosis of COVID-19 subjects that have undergone routine pathological blood tests.

This study proposes a Double Layer Stacked Ensemble Machine Learning model that can classify individuals, whether they are infected with nCoV or not, with high accuracy. Despite focusing on higher accuracy, this study also gave equal importance achieving high precision and recall so that no individual is misclassified. The unconventionality of this study is that the research has improved the overall performance metrics of the proposed model by analyzing consequences of misclassification based on different Basic Reproduction Numbers (R_0) related to the COVID-19 pandemic. The proposed model attempts to improve the COVID-19 patient classification ML algorithm so that a cheap and widely available COVID diagnostic method can be developed utilizing blood samples.

11.2 RELATED WORKS

AlJame et al. [3] created an ML model named ERLX for identifying COVID-19 from routine blood tests. In that experiment, two levels of the ML model were used in which the first level included Random Forest, Logistic Regression, and Extra Trees and an XGBoost classifier in another level for analyzing 18 features. The KNNImputer algorithm was used to manage null values in the dataset, iForest was used to eliminate outlier data, and the SMOTE technique was used to overcome imbalanced data; feature importance is described by using the SHAP system. In that study, there were 5644 sample data and 559 confirmed COVID-19 circumstances. The model obtained the result with average accuracy of 99%, AUC of 99%, a sensitivity of 98%, and a specificity of 99%.

Barbosa et al. [6] proposed a framework that helps Covid-19 determination dependent on blood testing. In this study, the laboratory parameters achieved 24 features from biochemical and the hemogram tests characterized to help clinical analysis. They used several types of machine learning techniques, namely Random Forest, SVM, Bayesian Network, and Naive Bayes and achieved accuracy of 95.2%, sensitivity of 96.8%, kappa index of 90.3%, specificity of 93.6%, and precision of 93.8%.

Kukar et al. [7] represented a method to predict COVID-19 infection via ML dependent on clinically accessible blood test results. Using the Random Forest algorithm

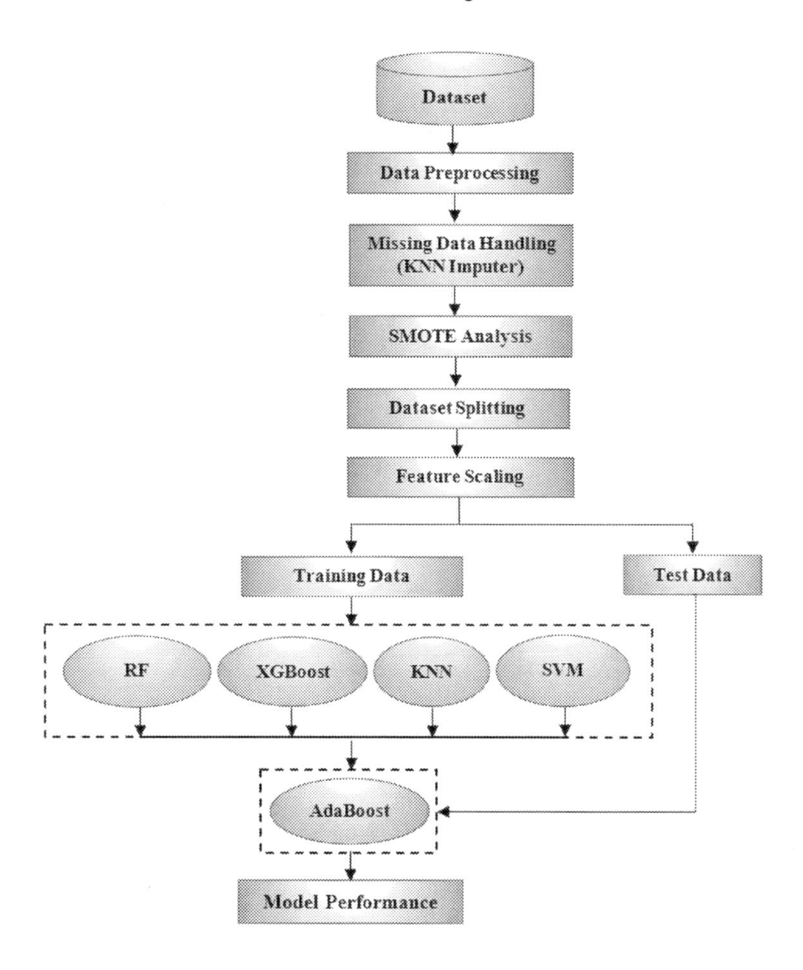

Figure 11.1 Block diagram describing the proposed method for the classification of COVID-19 based on blood sample data.

with 24 features, the COVID-19 screening achieved an in-depth clinical evaluation with accuracy of 91.67%.

11.3 METHODOLOGY

Blood tests in healthcare are performed in order to ascertain the biochemical and physiological state of a person. The whole dataset containing various attributes comprising hemocyte count and hematochemical profile obtained from two types of blood tests. A Complete Blood Count (CBC) and Basic Metabolic Panel are analyzed according to the block diagram portrayed in Figure 11.1.

11.3.1 Dataset Collection

The dataset used in this research was collected from Kaggle [18] and includes 5644 samples with 111 attributes. Among these, 603 samples and only 25 blood related attributes were selected for this study. The attributes are: Hematocrit,

Hemoglobin, Platelets, MPV, RBC, Lymphocytes, MCHC, Leukocytes, Basophils, MCH, Eosinophils, MCV, Monocytes, RDW, Neutrophils, Potassium, Creatinine, Sodium, Aspartate, Transaminase, INR, Albumin, Alanine, Transaminase, and C Reactive Protein.

11.3.2 Data Pre-processing

The following pre-processing procedures were performed in order to prepare the dataset for further analysis:

11.3.2.1 Missing Data Handling

Several techniques are available to counter missing values in a dataset. In this research work, the KNN imputer [3] has been applied with $k = 5$ neighbors to eliminate missing data contained in the dataset.

11.3.2.2 SMOTE Analysis

An imbalanced dataset makes the classifier model highly biased towards the high-frequency target class. To balance data, a well-known oversampling technique named Synthetic Minority Over-sampling Technique (SMOTE) [9] has been used in this study (Figure 11.2), which does data augmentation by grabbing the concept of k-nearest neighbors. Augmented data is intelligently added to be used in the classification problem to improve the data distribution. SMOTE does not augment data freely but similar to the existing data.

11.3.2.3 Data Splitting

The dataset was split into $70:30$ ratio as training and test sets. The stacked ensemble model was trained using 70% training data and an overall system evaluation was performed using 30% test data.

11.3.2.4 Feature Scaling

In this study, standardization has been used as the feature scaling method. The extremes are located near the average by a unit standard deviation in such a scaling process. This senses that the feature's average turns into zero, then the distribution takes a unit standard deviation after using this technique. [10]

11.3.3 Stacked Ensemble Machine Learning

This technique has widely been used since its invention, in which the 1st level learners are trained and utilized to make a primary prediction. The 1st level predictions are combined as the training dataset for a stacked 2nd level learner called the meta-learner and the test dataset is then fed to the 2nd level learner in order to make the final prediction. [11]

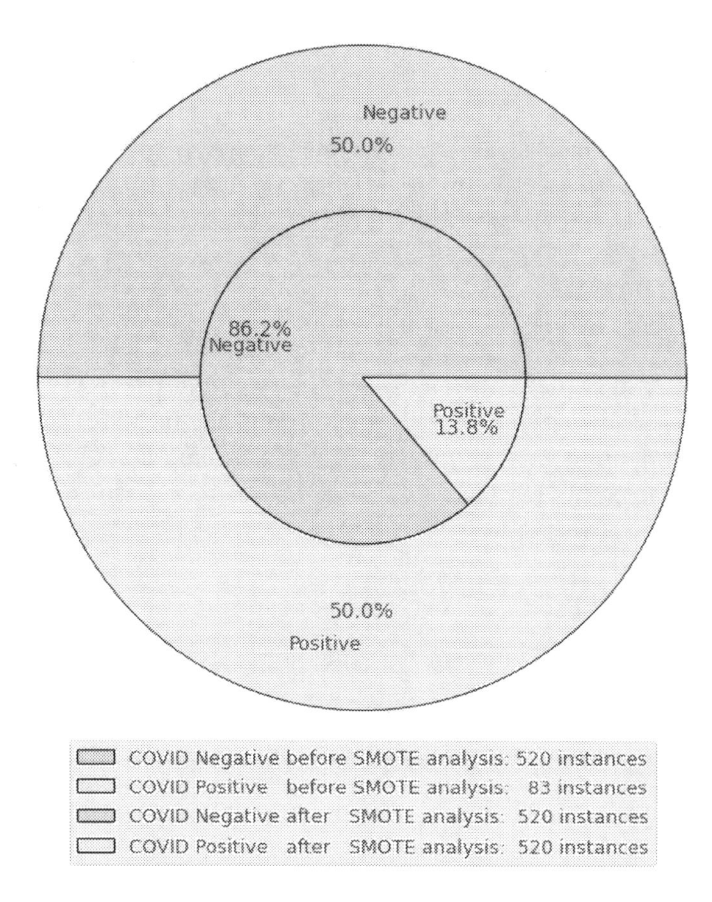

Figure 11.2 Data distribution before (inner-ring) and after (outer-ring) SMOTE analysis.

11.3.4 Machine Learning Algorithms

This study maneuvers double layer stacked ensemble machine learning in order to precisely classify subjects under Severe Acute Respiratory Syndrome (SARS) test caused by the SARS-CoV-2 strain of coronavirus as COVID positive or negative based on the hematological profile containing 25 attributes of 1040 subjects. Four algorithms (K-Nearest Neighbors, Support Vector Machine, Random Forest, XGBoost) were chosen for the 1st layer and AdaBoost was chosen as the metalearner for the 2nd layer of the proposed model based on high accuracy, precision, recall and F1-score.

11.3.4.1 K-Nearest Neighbors (KNN)

This algorithm measures the distance of the test datapoint from the training data-points and classifies the test datapoint according to the class of the k-closest training data-points in the neighborhood [12].

The hyperparameters this study used for KNN: **n_neighbors** = 3, **metric** = *minkowski*, **p** = 2.

11.3.4.2 Support Vector Machine (SVM)

This algorithm draws the best decision boundary called the 'hyperplane' at maximum distances from the target class in n-dimensional space so that new data points can certainly be categorized into the right classification later on [12].

The hyperparameters this study used for SVM: **kernel** = *rbf*, **random_state** = 0.

11.3.4.3 Random Forest (RF)

To overcome the pitfalls (data overfitting) of the Decision Tree (DT), RF produces a series of Decision Trees that have been trained using the 'bagging' technique wherein the general result is built from a multitude of DT learning models [12].

The hyperparameters this study used for RF: **n_estimators** = 49, **criterion** = *entropy*, **random_state** = 0.

11.3.4.4 XGBoost (XGB)

It's an optimized distributed gradient boosting library that utilizes ensemble learning techniques. Parallel and distributed computing from an ensemble of various learning algorithms makes it faster and efficient [13].

The hyperparameters this study used for XGB: **n_estimators** = 83, **max_depth** = 12, **subsample** = 0.7.

11.3.4.5 AdaBoost (AdB)

Boosting algorithms deal with bias-variance trade-off while bagging algorithms control variance only. The tweak behind improved accuracy and performance of AdB is the combination of the weighted sum of a few weak-learning classifiers that were misclassifying some instances when implied individually [13].

The hyperparameters this study used for AdB: **n_estimators** = 67, **learning_rate** = 1.

11.3.5 Compute Statistical Metrics

The following statistical metrics were used to measure the performance of this proposed work: accuracy, precision, recall, and f1 score. Individual output is calculated using confusion matrix values such as true negative (TN), true positive (TP), false negative (FN), and false positive (FP) [14].

11.4 OUTCOMES

The performance metrics of the detailed study are tabulated in Table 11.1 and graphically rendered in Figure 11.3.

At the 1st level classifier, analysis was carried out using different machine learning algorithms (LR, KNN, DT, ANN, SVM, XGB, RF, NB and AdB) in an effort to find the algorithms best suited for the 1st layer of the proposed model. Based on accuracy,

Table 11.1 Performance Metrics of the Proposed Stacked Ensemble Model.

	Layer-1				Layer-2
Performance Metrics	KNN	SVM	XGB	RF	AdB
Accuracy	84.62%	92.95%	95.19%	97.12%	100%
Precision	98.73%	96.84%	97.47%	99.37%	100%
Recall	77.23%	90.00%	93.33%	95.15%	100%
F1 Score	86.67%	93.29%	95.36%	97.21%	100%

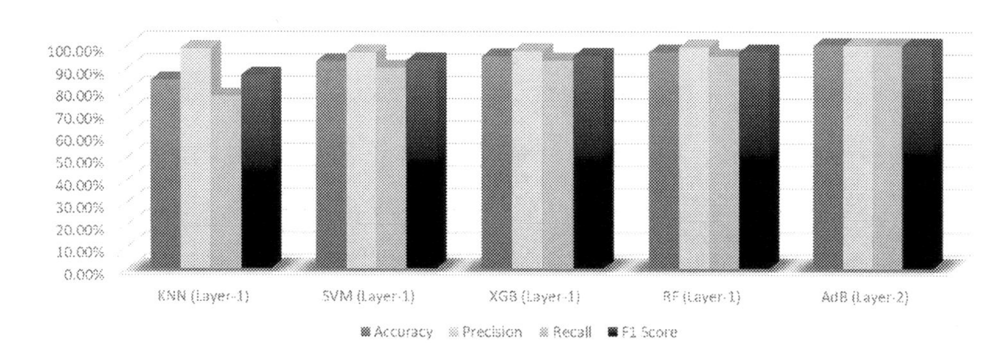

Figure 11.3 Performance Metrics (Graphical Visualization) of the Proposed Stacked Ensemble Model.

precision, recall, F1-score and giving priority to precision, four algorithms were chosen for the 1st layer of the proposed model, which are RF, XGB, SVM and KNN.

RF, XGB, SVM and KNN algorithms have accuracy of 84.62%, 92.95%, 95.19% and 97.12%; precision of 98.73%, 96.84%, 97.47% and 99.37%; recall of 77.23%, 90%, 93.33% and 95.15%, and F1-score of 86.67%, 93.29%, 95.36% and 97.21% respectively.

It's worth noting that despite better performance, none of the algorithms in the 1st layer are reasonably precise enough to curb the community transmission of COVID-19.

For illustration of the associated risks, let's consider the confusion matrix (Figure 11.4(a)) of the best performing algorithm (RF) of the 1st layer.

In the 1st layer, for RF, the corresponding precision is 99.37% and recall is 95.15%. Note that precision reveals the caliber of a model to classify positive values correctly and recall tells how often the model actually predicts the correct positive values' [14].

It's clear from the confusion matrix of Figure 11.4(a) that the number of people who are actually COVID negative but misclassified as positive (False Positive) is 8. If these FP individuals are now kept in Quarantine/Isolation, they will lose working hours ultimately reducing their monthly income and the country's GDP growth. In the worst-case scenario, if they are medicated for COVID-19, it's left for the reader to think about the consequence of the wrong medication if the individuals had other chronic diseases.

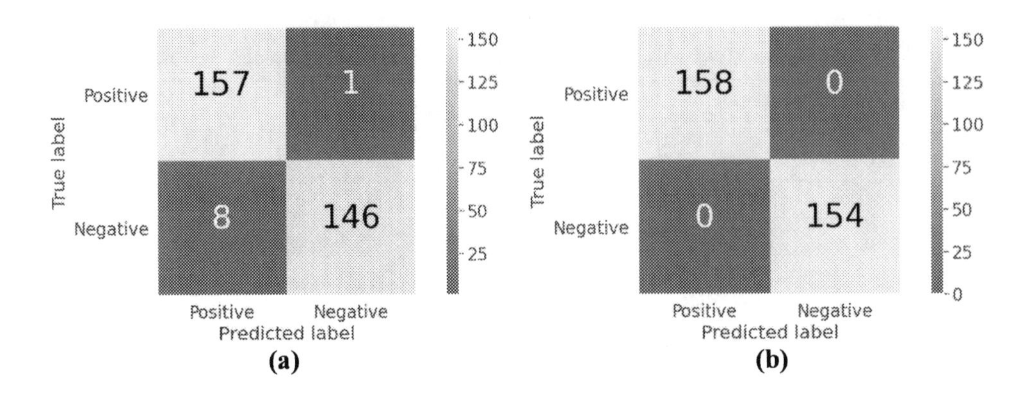

Figure 11.4 Confusion matrices of the best algorithm (a) Random Forest in 1st layer (b) AdaBoost in 2nd layer.

Again, the dangerous and risky scenario of the confusion matrix of Figure 11.4(a) is the number of people who are actually COVID positive but misclassified as negative (False Negative) is 1. This 'False Negative' classification is of utmost importance because these misclassified people are enough to spread COVID through community transmission if they move freely because they received a negative COVID test which is wrong.

According to the Simple Infectious Disease Model [15], after the incubation period of τ days, if an infected person infects exactly R_0 new susceptibles, then the number of new infected individuals only from him/her is given as:

$$n_E(t) = n_E(0)R_0^{t/\tau} \tag{11.1}$$

Considering 1 initial infected (misclassified as negative) person and 7 days incubation period, the equation is graphically visualized in Figure 11.5(a) describing one month projection of new infection cases by only 1 person due to community transmission of COVID for various R_0 values. It will seldom be possible to curb the pandemic if newly infected persons spread the disease again like Figure 11.5(c).

Now, to quantify the risk regarding infectious disease spread like Figure 11.5(c), let an area having a population of 20,000, where initial infectives (misclassified as COVID negative), $I_0 = 1$ and remaining susceptible, $S_0 = 19{,}999$. According to the simple SIR model [16] for COVID-19:

Contact ratio, $q = \frac{R_0}{S_0}$ and maximum number of people that will be infected, $I_{max} = I_0 + S_0 - \frac{1}{q}(1 + ln(qS_0))$. Using the recent COVID Reproduction Number (R_0) of Bangladesh Figure 11.5(d) [17] and the equations above, Figure 11.5(e) depicts an approximation of the maximum number of new infection cases.

This study used a relatively small test set in the 1st layer, where the best algorithm gave FN $I_0 = 1$. In case of larger datasets of a bigger area, this algorithm would contribute higher FN I_0 to the equation of I_{max} and consequently Figure 11.5(e) would become worse.

If the machine learning algorithms with high accuracy but unknown/less precision and recall of the reviewed literature (Table 11.2) keep performing FN misclassification

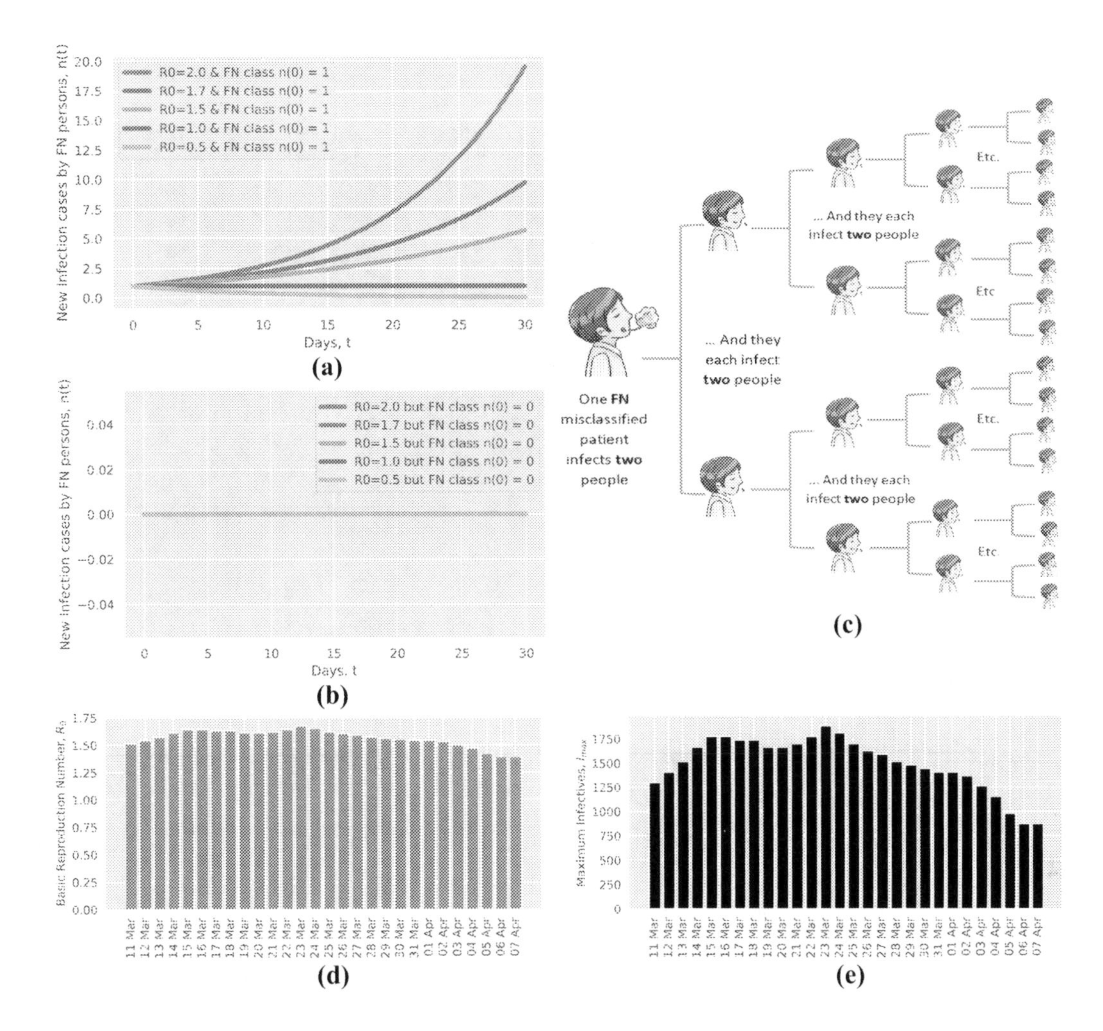

Figure 11.5 (a) R-curve: Effect of FN misclassification based on different R_0 values according to the Simple Infectious Disease Model. (b) R-curve remains at zero if the individuals were properly classified. (c) Rapid contagious nature of infectious disease if $R_0 = 2$. (d) COVID-19 R_0 values of Bangladesh during 11 March 2021 to 7 April 2021. (e) Using R_0 values, rough idea of new infection cases as per simple SIR model.

due to less precision and recall, Figure 11.5 clearly says, the *pandemic* will remain a *pandemic* due to community transmission from the FN class.

Therefore, the novelty of the idea and algorithm of this study is, it alleviates the risk factor associated with less precision and recall using the R-curve plot and I_{max} plot from the SIR model using simplified-generalized equations. It is imperative to resolve the FN and FP classifications, for which this research work emphasized improving precision and recall along with achieving high accuracy by feeding the output of the 4 selected algorithms of the 1st layer of the proposed model to the 2nd (final) layer as input.

In the 2nd (final) layer, called the meta-learner of the stacked ensemble model, after applying the AdaBoost algorithm, the model achieved improved performance

Table 11.2 Comparison of the proposed model with other existing models.

Ref	#Feat	Algorithms Used	Acc	Pre	Rec	F1
[3]	18	ExtraTrees, RF, Logistic Regression, XGB	99.88%	-	-	-
[6]	24	SVM, MPL, RF, Naive Bayes, Bayesian network	95.16%	-	-	93.80%
[7]	24	Random Forest	91.67%	-	-	-
[4]	14	CR	84.21%	83.70%	84.20%	83.70%
[5]	15	SVM, RF, DT	84.00%	-	-	92.00%
[8]	32	SVM	81.48%	-	-	-
Proposed Model	25	Stacked Ensemble Machine Learning	100%	100%	100%	100%

Note: Ref = References, #Feat = Number of Features, Acc = Accuracy, Pre = Precision, Rec = Recall, F1 = F1-Score

(Table 11.1 and Figure 11.3) where the AdB algorithm outperformed others, achieving 100% on all the performance metrics.

Visually, the confusion matrix, Figure 11.4(b), of the best algorithms (AdB) of the 2nd layer shows that there is no FN and FP classification.

Hence, according to the R-curve of Figure 11.5(b), the new infection cases by FN class will be zero irrespective of different R_0 values. Consequently, if the true positive individuals are held in isolation and all social norms are maintained, gradually the R_0 number will go below 1 and new infection cases will reduce like Figure 11.5(a) where $R_0 < 1$.

Also due to the zero FP class, the true COVID negative individuals will get relief from: being misclassified, losing working hours and from taking incorrect medication in the worst cases.

Thus, the proposed highly accurate and risk-free Stacked Ensemble Machine Learning model is capable of classifying individuals as COVID-19 +ve or -ve based on hematological profile of their blood samples.

11.5 CONCLUSION

To combat the pandemic, it is imperative to ramp up the number of tests conducted in a country with high accuracy and precision for reliable screening of infected patients in a short time. To supplement the conventional time-consuming and scarce RT-PCR test, this study proposes a risk-free Stacked Ensemble Machine Learning model to diagnose COVID-19 patients with relatively higher accuracy and precision from their hematological data obtained from routine blood tests.

This study characterized the limitations and risks associated with the models having unknown or less precision and recall in case of COVID-19 diagnosis. It also exploits the R-curve and maximum number of infections from the simple SIR infectious disease model generalized for COVID-19 and hence harnessed the versatility and robustness of the stacked ensemble model to resolve the problem. In the future, further development will address some challenges including use of a huge, diverse, high-quality dataset, rigorous public testing and validation, incorporation of more attributes, and multiclass classification rather than the binary classification performed in this work.

11.6 SUPPLEMENTARY WEBLINK

The video presentation briefly describing this chapter can be accessed from: https://youtu.be/Ci8dznDadJ4

Bibliography

[1] Schiffmann, A. 2021. *Coronavirus Dashboard.* Accessed 27 July, 2021. https://ncov2019.live/.

[2] Vogels, C.B., A.F. Brito, A.L. Wyllie, J.R. Fauver, I.M. Ott, C.C. Kalinich, M.E. Petrone, et al. 2020. Analytical Sensitivity and Efficiency Comparisons of Sars-Cov-2 Rt–Qpcr Primer–Probe Sets. *Nature Microbiology* 5, no. 10: 1299–1305.

[3] AlJame, M., I. Ahmad, A. Imtiaz, and A. Mohammed. 2020. Ensemble Learning Model for DIAGNOSING COVID-19 from Routine Blood Tests. *Informatics in Medicine Unlocked* 21: 100449.

[4] Arpaci, I., S. Huang, M. Al-Emran, M.N. Al-Kabi, and M. Peng. 2021. Predicting the COVID-19 Infection with FOURTEEN Clinical Features Using Machine Learning Classification Algorithms. *Multimedia Tools and Applications* 80, no. 8: 11943–11957.

[5] Bao, F.S., Youbiao He, J. Liu, Yuanfang Chen, Q. Li, Christina Zhang, Lei Han, B. Zhu, Y. Ge, Shi Chen, M. Xu and L. Ouyang. 2020. Triaging moderate COVID-19 and other viral pneumonias from routine blood tests. *ArXiv* abs/2005.06546 (2020): n. pag.

[6] de Freitas Barbosa, V.A., J.C. Gomes, M.A. de Santana, J.E. Albuquerque, R.G. de Souza, R.E. de Souza, and W.P. dos Santos. 2021. Heg.IA: An Intelligent System to Support Diagnosis of Covid-19 Based on Blood Tests. *Research on Biomedical Engineering*: 1–18.

[7] Kukar, M., G. Gunčar, T. Vovko, S. Podnar, P. Černelč, M. Brvar, M. Zalaznik, M. Notar, S. Moškon, and M. Notar. 2021. COVID-19 Diagnosis by Routine Blood Tests Using Machine Learning. *Scientific Reports* 11, no. 1.

[8] Yao, H., N. Zhang, R. Zhang, M. Duan, T. Xie, J. Pan, E. Peng, et al. 2020. Severity Detection for THE Coronavirus DISEASE 2019 (COVID-19) Patients Using a Machine Learning Model Based on the Blood and Urine Tests. *Frontiers in Cell and Developmental Biology* 8: 683.

[9] Akter, L., and N. Akhter. 2020. Detection of Ovarian MALIGNANCY from Combination of Ca125 in Blood AND TVUS Using Machine Learning. *Advances in Intelligent Systems and Computing*: 279–289.

[10] Pedregosa, F., Varoquaux, G., Gramfort, A., Michel, V., Thirion, B., Grisel, O. et al. 2021. Sklearn.Preprocessing.Standardscaler Scikit-Learn 0.24.2 Documentation. *Journal of Machine Learning Research* https://scikit-learn.org/stable/modules/generated/sklearn.preprocessing.StandardScaler.html.

[11] Geron, A. 2019. Chapter 7: Ensemble Learning and Random Forests. Essay. In *Hands-on Machine Learning with Scikit-Learn and amp; TENSORFLOW: Concepts, Tools, and Techniques to Build Intelligent Systems*, 210–213. 2nd ed. Sebastopol, CA: O'Reilly Media, Inc.

[12] Shams, A.B., E. Hoque Apu, A. Rahman, M.M. Sarker Raihan, N. Siddika, R.B. Preo, M.R. Hussein, S. Mostari, and R. Kabir. 2021. Web Search Engine Misinformation Notifier Extension (Seminext): A Machine Learning Based Approach during Covid-19 Pandemic. *Healthcare* 9, no. 2: 156.

[13] Gomez-Rios, A., J. Luengo, and F. Herrera. 2017. A Study on the Noise Label Influence in Boosting ALGORITHMS: ADABOOST, GBM and XGBoost. *Lecture Notes in Computer Science*: 268–280.

[14] Raihan, M.M., A.B. Shams, and R.B. Preo. 2020. Multi-Class Electrogastrogram (EGG) Signal Classification Using Machine Learning Algorithms. *2020 23rd International Conference on Computer and Information Technology (ICCIT)*: 1–6.

[15] Gog, J., R. Thomas, and M. Freiberger. 2021. The Growth Rate of Covid-19. *Plus.maths.org*. https://plus.maths.org/content/epidemic-growth-rate.

[16] Murray, J.D. 2001. 10.2 Simple Epidemic Models and Practical Applications. Essay. *In Mathematical Biology: An Introduction with 189 Illustrations*, 17:323–323. 3rd ed. New York: Springer.

[17] Arroyo-Marioli, F., F. Bullano, S. Kucinskas, and C. Rondón-Moreno. 2021. Tracking R of Covid-19: A New Real-Time Estimation Using the Kalman Filter. *PLOS ONE* 16, no. 1.

[18] Einstein Data4u, 2020. Diagnosis of COVID-19 and Its Clinical Spectrum. https://www.kaggle.com/einsteindata4u/covid19.

CHAPTER 12

A Transfer Learning Approach to Recognize Pedestrian Attributes

Saadman Sakib

Chittagong University of Engineering & Technology (CUET), Chattogram, Bangladesh

Anik Sen

Chittagong University of Engineering & Technology (CUET), Chattogram, Bangladesh

Kaushik Deb

Chittagong University of Engineering & Technology (CUET), Chattogram, Bangladesh

CONTENTS

PEdestrian attribute recognition has recently created a significant impact because of its soft bio-metric properties for recognizing individuals. The popularity of the pedestrian attribute recognition task is attributable to deep learning, especially the convolutional neural network. This chapter presents a transfer learning approach for the pedestrian attribute recognition task because of its high performance and low training cost. The Mask RCNN object detector extracts images of isolated pedestrians. After that, the preprocessed images are passed to different CNN architectures, i.e., Inception ResNet v2, Xception, and ResNet 101 v2, to extract the spatial features. Experiments reveal that the Xception architecture outperforms the competition with a 90.33% accuracy rate. Following that, some experiments on the Xception architecture are carried out by freezing layers, except the last 4, 8, 12, 16, 20, all, and no layers (excluding the fully connected layers). For the RAP v2 dataset, experimental results show that freezing all layers except the last 16 layers provides the best accuracy, 92.52%, outperforming existing methods in terms of accuracy.

12.1 INTRODUCTION

The modern world produces vast quantities of data regularly as a result of technological advances. In the field of computer vision, image data makes an important contribution. Image data and advanced techniques are necessary for performing image recognition, image classification, or object detection. Image recognition aims to automatically recognize an object or a group of objects without the need for human interference. This process is focused on the semantic content of an image.

Pedestrian attribute recognition, in this context, refers to the classification of pedestrian attributes in surveillance scenarios based on semantic features in [1]. Pedestrian images have spatial correlations that are used to identify pedestrian attributes. Examples of pedestrian attributes are gender, age, upper body clothing, lower body clothing, shoes, etc. Pedestrian attributes can be used in pedestrian detection [2]. Also, it can be used in a person re-identification task too [3].

An essential method for extracting features from images is the Convolutional Neural Network. Recently, several intriguing deep CNN architectures have been published as proposed in [4]. Transfer learning techniques have gained popularity. This technique requires less training data as well as fewer epochs for a new task. The multi-label classification issue of pedestrian attribute recognition has been widely used in the retrieval and re-identification of individuals. Previous works [5] also introduced pre-trained architectures. Inspired by this, we experimented with different pre-trained architectures. The pedestrian attribute recognition task is shown in Figure 12.1. In a nutshell, the research contribution can be summarized as follows:

- The transfer learning-based architecture Xception which can recognize pedestrian attributes, outperformed existing state-of-the-art methods with 92.52% accuracy.

The rest of the chapter is organized as follows. Section 12.2 describes the related works for our proposed method. Section 12.3.1 discusses the overview of the proposed method. Additionally, Section 12.3.2 describes the Mask RCNN object detector. In Section 12.3.3, the preprocessing techniques are discussed. Then spatial feature extraction and transfer learning approaches are described in Sections 12.3.4, and 12.3.5 respectively. Section 12.3.6 describes the classifier used in the proposed method. Dataset description is explained in Section 12.4.1. Also, results and discussions are shown in Section 12.4.3. Finally, Section 12.5 outlines the conclusion and future works for the proposed method.

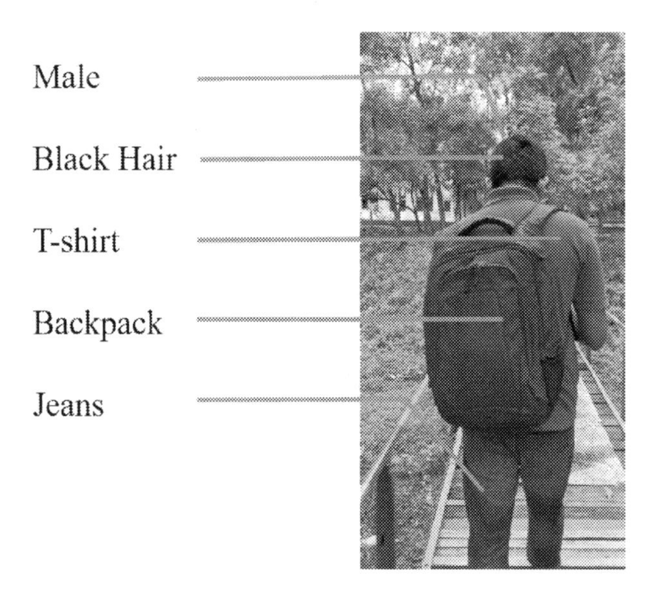

Figure 12.1 Example of the pedestrian attributes.

12.2 RELATED WORKS

Pedestrian attribute recognition is a multi-label classification problem. When computational cost was a significant issue for using deep learning techniques, handcrafted feature-based methods were used to perform image processing tasks. Nowadays, with the advancement of computers, the computation issue is almost solved. Earlier studies on this topic did not gain better results. Reference [6] showed how the AdaBoost method is used to train a discriminative classification system instead of constructing a feature by hand to address the issue. [7] developed an Ensemble RankSVM to address the scalability issue that plagues existing SVM-based ranking techniques. This new model requires much less memory, allowing it to be far more versatile while maintaining excellent performance.

In [8, 9] a novel re-identification approach was proposed that learns a range and weighting of mid-level semantic attributes to identify individuals. The model learns an attribute-centric, parts-based feature representation in particular. They also demonstrated how mid-level semantic attributes for individual definitions could be computed.

Deep learning methods, such as deep Convolutional Neural Networks (CNN), on the other hand, have shown superior results in a variety of computer vision tasks. DeepSAR (where the attributes are considered independent) and DeepMAR (where the attributes are regarded as dependent) are two CNN-based attribute recognition methods proposed in [10]. They outperformed the MRFr2 method, which was state-of-the-art at the time.

Reference [11] suggested a method for jointly training a CNN model for all attributes that can reap the benefits of attribute dependencies, using only the image as input and no other external pose, part, or context information. Their multidisciplinary approach is unique.

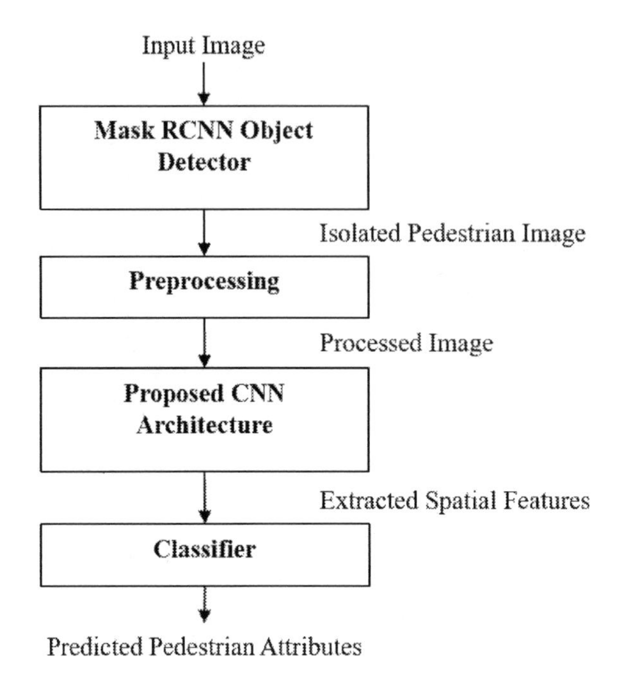

Figure 12.2 Steps of proposed method for pedestrian attribute recognition.

In [5] a grouping approach was proposed based on a fine-tuned VGG-16 structure, in which the attributes are first grouped and then fed into a pre-trained VGG-16 model with minor modifications for improved performance. A custom-tailored CNN network for a particular task will outperform a pre-trained network like AlexNet, or ResNet according to [12]. As the transfer learning technique has been used in previous works, we also introduced the transfer learning technique in our proposed work.

12.3 METHODOLOGY

The proposed method for the pedestrian attribute recognition task is discussed in this section.

12.3.1 Overview

The classification of an object in an image is known as image recognition. Images include spatial features that are used to classify the object. Figure 12.2 depicts the steps of the proposed method for pedestrian attribute recognition. At first, the input image is passed to the Mask RCNN object detector to get the isolated pedestrians. After that, the isolated pedestrian image is resized since the CNN model works on a uniform image size. Scaling is applied to an image to minimize computational complexity. The image augmentation phase is in charge of introducing variance into the CNN model for training. In our proposed method, the preprocessing techniques are resizing, scaling, and augmentation. It aids the CNN model in learning more effectively. After that, the CNN architecture uses the transformed image to extract

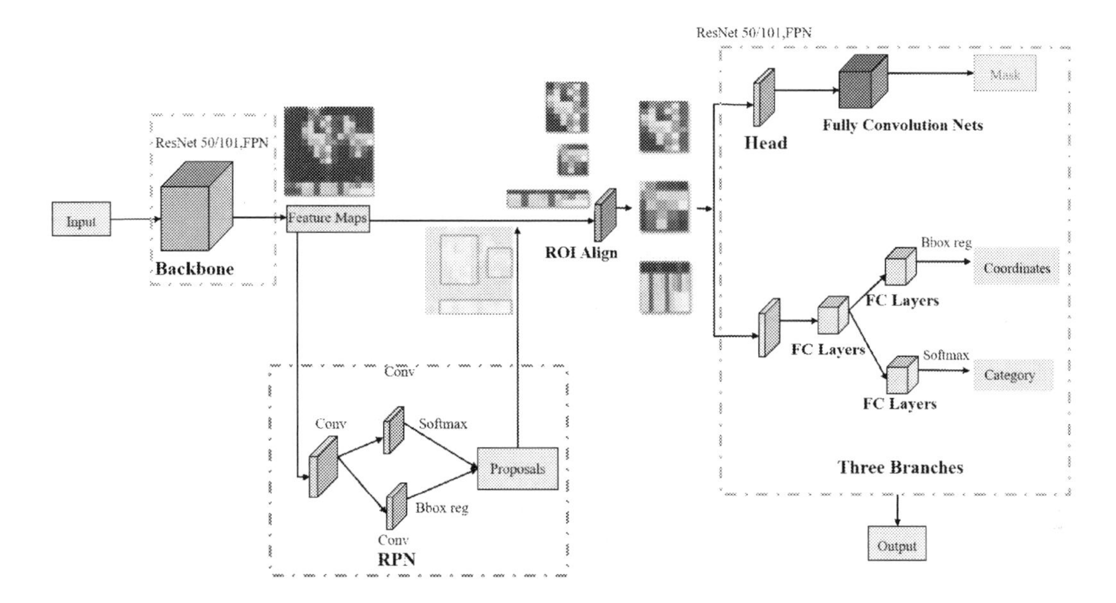

Figure 12.3 Mask RCNN Object Detector Framework.

spatial attributes. The classifier is a neural network that uses the extracted features as input and outputs a prediction of pedestrian attributes.

12.3.2 Mask RCNN Object Detector

Pedestrian identification has advanced quickly in recent years, and it is now being used more widely. As described in [13], many applications depend on pedestrian detection, including smart vehicles, object tracking, robotics, and video surveillance. Pedestrian detection has received a lot of attention in the computer vision area because of its importance. Figure 12.3 shows the basic overview of the Mask RCNN object detector framework.

The blocks in Figure 12.3 are explained in the following:

- **Backbone:** The backbone on the Mask RCNN extracts features from the given input image. It can use ResNet 50 or 101 or any Feature Pyramid Network(FPN). In our proposed work, we have used ResNet 50 for the task.

- **RPN:** The extracted features are passed to a Region Proposal Network, which proposes regions where objects are present in an image. It performs some convolution operations and predicts the bounding box and the probability of an object's presence.

- **ROI Align:** Then the Region of Interest Align Network works to give three types of output, i.e., the mask of the objects, the bounding box of the objects, and the label or class name of the objects.

12.3.3 Preprocessing

Image preprocessing is a technique for applying operations to an image in order to improve it or extract useful information from it. In Deep Learning, image preprocessing is needed to take advantage of the processed image. It has been demonstrated that image preprocessing improves the efficiency of a convolutional neural network significantly.

Image resizing means changing the resolution of an image. The main purpose of this task is to reduce computational complexity. In the meantime, reducing the size of an image also reduces the quality. So considering these constraints, we resized all the RGB images into 150×150 resolution.

After resizing, the images are scaled. The RGB images have pixel values in the range [0,255]. With a large pixel value, the computational cost also increases for CNN. So to reduce the cost, all the pixel values are scaled in the range [0,1]. This is done by dividing the pixel values by 255, as 255 is the max pixel value in an RGB image.

The augmentation technique was applied to the processed image after resizing and scaling. We used augmentation during the training process in this study. There are several batches of images in each epoch. Different augmentation techniques are used at random in each batch. We have, however, considered five different transformation methods, i.e., width shift, height shift, rotation, shearing, and horizontal flip.

12.3.4 Spatial Feature Extraction

An image has distinct characteristics. Features are the pixels in a picture that are responsible for the uniqueness of an entity in the image. Spatial feature extraction is the process of extracting features from an image that will later be used to describe or classify an entity. A convolutional neural network, which performs a series of operations, is used to extract the spatial features of an image. Convolution is followed by an activation function, which is followed by pooling.

12.3.5 Transfer Learning Approach

Deep Learning techniques require a large amount of data to learn. Sometimes this large amount is not available. Also, making a Deep Learning Network learn from scratch is very time-consuming. That's where the transfer learning technique comes in. Earlier, [14] proposed a transfer learning technique to detect pedestrians. They did not introduce a CNN in their work. After that, more powerful CNN architectures with a transfer learning approach have been used to detect pedestrians. [15] proposed the Mask RCNN object instance segmentation architecture, which is very effective for object detection tasks too. Also, [16] proposed the YOLO (You Only Look Once) architecture, which is efficient for object detection tasks. [17] demonstrated a comparison among these powerful CNN architectures. They showed that YOLO is faster than Mask RCNN. However, in terms of average precision, Mask RCNN outperforms all other architectures by 47.3%. We have used the Mask RCNN object detection architecture for better object detection purposes in our proposed method. Also, we only used transfer learning to recognize the pedestrian attributes, which is the next

Table 12.1 Comparison of pre-trained models on RAP v2 dataset.

Pre-trained Models	Trainable Parameters	Nodes in Dense Layer	Test Accuracy (%)
Inception ResNet V2	55.9M		90.37
ResNet 101 V2	44.7M	1024	91.01
Xception	22.9M		**92.26**

step of pedestrian detection.The transfer learning technique transfers the knowledge from an already trained dataset. This knowledge helps detect edges or other complex features. Training a network from scratch can also obtain this knowledge, but this will only increase the cost. So to reduce time and cost, the transfer learning technique is chosen in our proposed architecture. Because [5] used the transfer learning technique, we also experimented with different pre-trained architectures. Initially, we obtained the Xception architecture as the best-performing architecture. Other works did not perform such vivid experiments. After that, we fine-tuned the Xception architecture by freezing all layers except the last 4, 8, 12, 16, 20, all, and no layers (excluding the fully connected layers). This experiment is unique in this field.

- **Comparison of Pre-trained Models:** Before the transfer learning approach is used in our proposed method, some experiments are performed on pre-trained architectures, i.e., Inception ResNet v2, Xception, and ResNet 101 v2, to select the best performing architecture. The architectures are chosen based on their performance on the IMAGENET dataset. A comparison of the pre-trained models on the RAP v2 dataset is shown in Table 12.1.

  From Table 12.1 we can deduce that the Xception architecture performs better than other architectures. The Xception architecture obtained 92.26% accuracy on the RAP v2 dataset.

- **Proposed CNN Architecture:** The IMAGENET dataset is used to pre-train the Xception architecture. It obtained 94.5% on top 5 accuracy as shown in [18]. The IMAGENET dataset has variety in it. It has over 14M images and 1000 classes. The architecture can take an input image of dimension $299 \times 299 \times 3$. However, this large input dimension will increase the computational cost. So we used $150 \times 150 \times 3$ as the dimension of the input image. The architecture of Xception is shown in Figures 12.4 and 12.5.

  The details of each block of Figure 12.4 is shown in Figure 12.5. The blocks from Figure 12.4 are explained in the following:

  - **Conv** 3×3**:** This is the first block of the Xception architecture. From Figure 12.4 we can see that it has been repeated two times with a 3×3 convolution operation. Thus after this block, the feature map becomes smaller. In general, this block reduces the feature map.

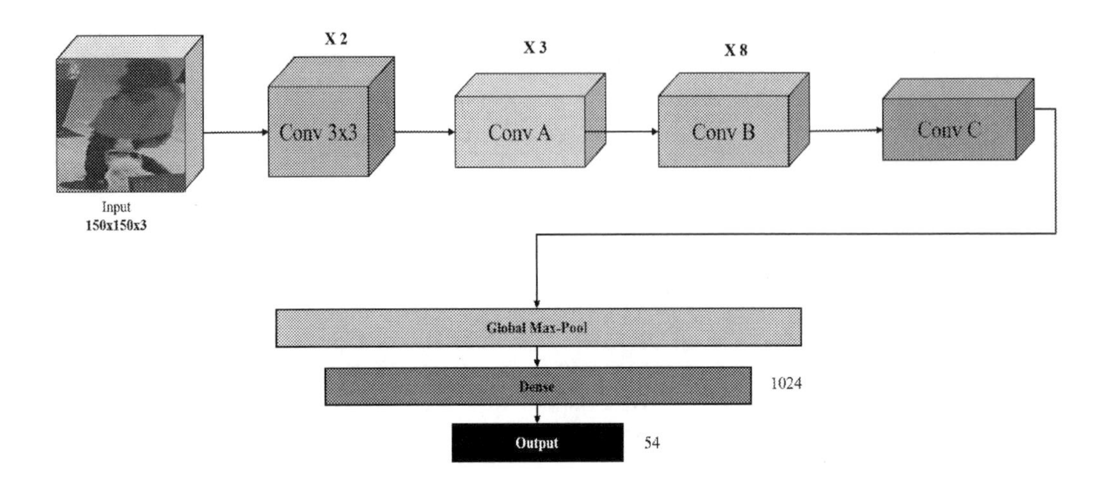

Figure 12.4 Overview of Xception architecture.

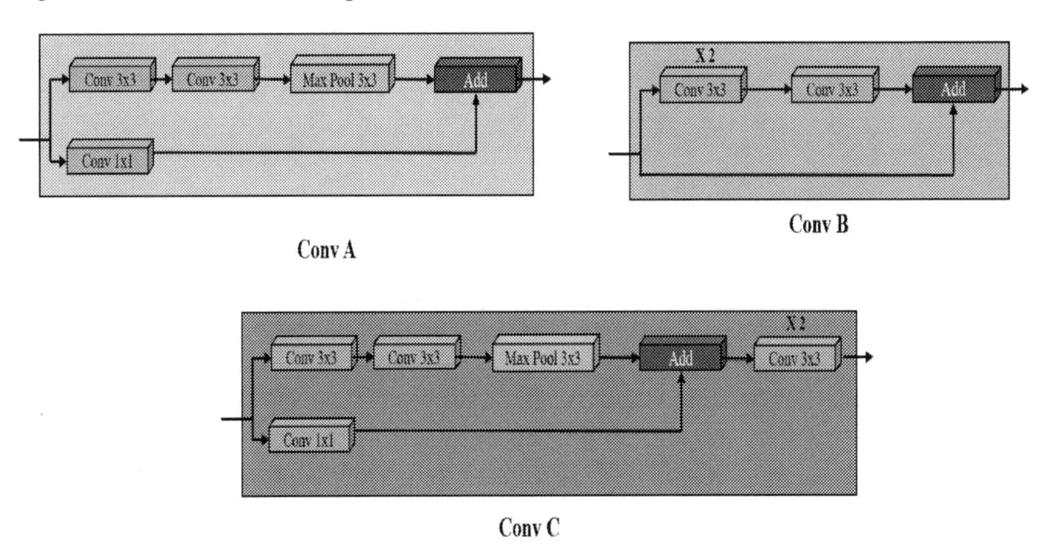

Figure 12.5 Details of the blocks of the Xception architecture.

- **Conv A:** This block takes advantage of residual connection, which helps the network learn from previous layers and inception modules to reduce computation costs. A 1×1 convolution operation is performed on every channel, then a 3×3 convolution operation is performed on each output in Figure 12.5.
- **Conv B:** This block has only a residual connection to learn from previous layers. The convolution 3×3 operation is performed in this block as shown in Figure 12.5.
- **Conv C:** Similar to the Conv A block with more convolution 3×3 operations at the end of this block. The purpose of this block is to reduce the

feature map size. Convolution and max pool operations are used for this reason, as shown in Figure 12.5.

After the last Conv C block, the extracted features are passed to the Global Max Pool layer. This layer performs the same operation as the Max pool, except the window size is the same as the input width and height. So it reduces the number of features and takes the most dominant feature in a channel.

12.3.6 Classifier

After the spatial feature extraction from the CNN architecture, a classifier is used to predict the pedestrian attributes. As a neural network is used in the proposed architecture, the classifier has one dense layer. Dense layers are required to give efficiency to a neural network. As the problem grows more complex, the network should indeed learn complex non-linearity. The dense layer introduces non-linearity as the ReLU activation function is used in every node. In our case, we used 1024 neurons in the dense layer. Finally, an output layer of 54 nodes is chosen because our proposed method is a multi-label classification task with 54 binary attributes.

The sigmoid activation function is used in the output layer as it gives a probability between 0-1. If the probability is greater than 0.5, we considered the attribute positive, meaning the attribute is present in the pedestrian. Otherwise, we considered it as 0, which means absent. We can get the output from an output layer by the equation 12.1

$$Z = f(W.X + b), \tag{12.1}$$

where W and X are the weight matrix and input matrix, respectively. b is the bias of every node of the input layer. f represents the activation function, which is the sigmoid activation function in our case. Z is the output.

12.4 OUTCOMES

The trial was carried out on Google Colab, a freely accessible open cloud service operated by Google. A total of 12 GB of RAM was given, with 12-hour runtime. There was also 14 GB of Tesla K80 GPU memory available. The proposed method was built using Keras with Tensorflow 2.0 as the backend and Python 3.7 as the programming language.

12.4.1 Dataset Description

Many datasets are available for the purpose of recognizing pedestrian attributes. [19] proposed the Richly Annotated Pedestrian (RAP) dataset with a total of 41,585 pedestrian samples annotated with 72 attributes from real multi-camera surveillance scenarios. Later, [20] proposed the RAP dataset, which is a large-scale dataset that comprises 84,928 images with 72 categories of attributes and additional identifiers.

The resolution of the images in the dataset varies from 33×81 to 415×583. Twenty-five camera scenes from an indoor shopping mall captured the images. We

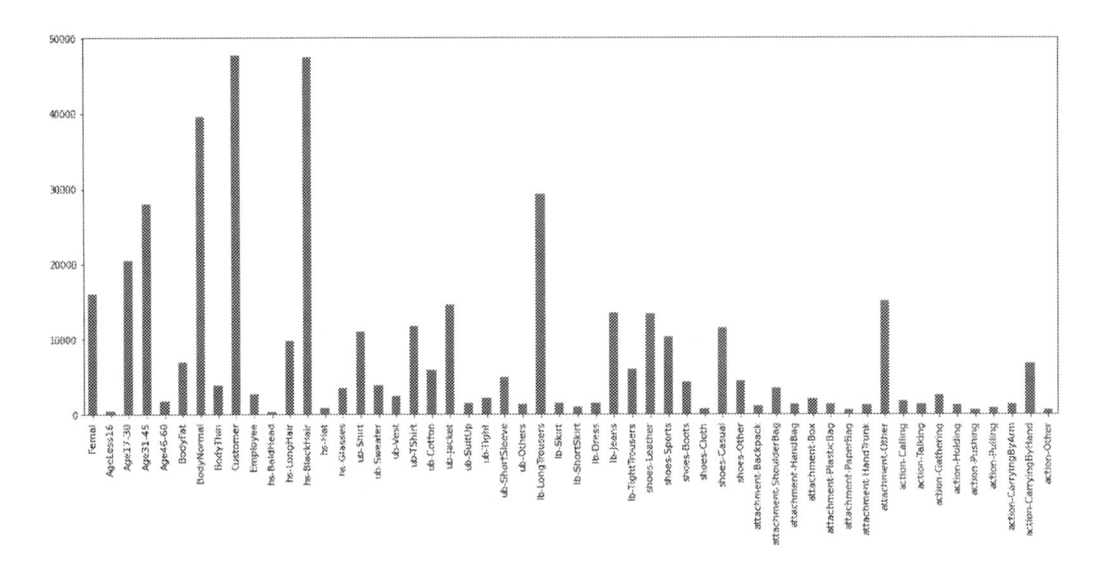

Figure 12.6 Distribution of positive samples for training data with 54 attributes.

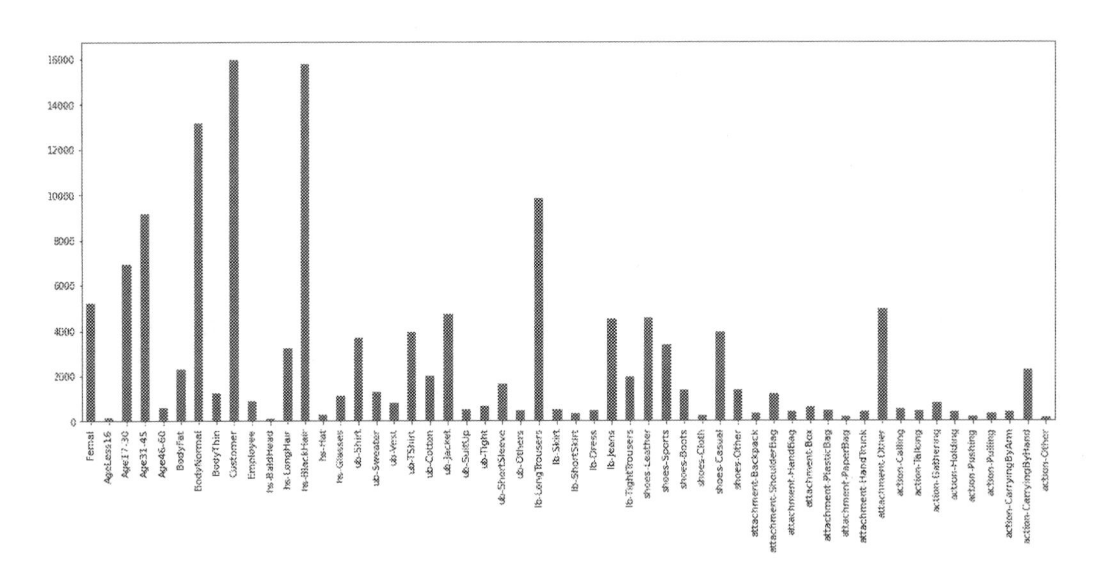

Figure 12.7 Distribution of positive samples for test data with 54 attributes.

used the ratio of (6: 2: 2) as stated in the last section. Here 60% data is used for training and 20% for testing and validation. The dataset is highly imbalanced. For consistency, the split for the dataset taken is the same for other previous work.

Fifty-four binary attributes are chosen for the proposed method. As the dataset is imbalanced, the distribution of positive samples for training, test and validation data are shown in Figures 12.6, 12.7, and 12.8, respectively.

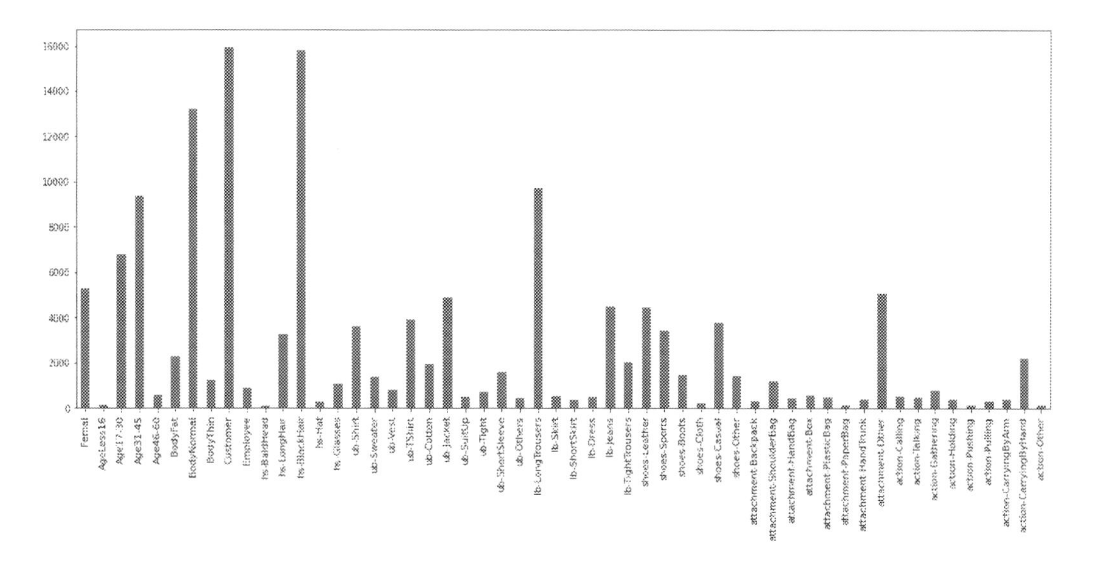

Figure 12.8 Distribution of positive samples for validation data with 54 attributes.

Table 12.2 Experiment on Xception architecture via transfer learning method.

Xception Models	Trainable Layers (Excluding FC Layer)	Nodes in Dense Layer	Trainable Parameters	Test Accuracy (%)
Model 1	None		2,155,574	90.33
Model 2	Last 4 layers		5,319,222	91.14
Model 3	Last 8 layers		6,906,422	92.06
Model 4	Last 12 layers	1024	8,405,966	92.36
Model 5	**Last 16 layers**		**8,943,958**	**92.52**
Model 6	Last 20 layers		9,481,950	91.64
Model 7	All		22,962,526	92.26

12.4.2 Experiments on the Proposed CNN Architecture

After selecting primarily Xception, we performed a vivid experiment on this architecture. In using the transfer learning method, we tried to freeze layers of the architecture. Here, freezing means the weights of the frozen layers will remain unchanged in the training phase. No data balancing techniques are applied in this experiment. The purpose is to increase the performance of the proposed architecture. After much experimentation, Table 12.2 was obtained.

From the Table 12.2, we can see that the accuracy is increased by training the last 16 layers of the Xception architecture. The obtained accuracy is 92.52%. For consistency, we used a dense layer of 1024 nodes in all cases and 40 epochs per experiment.

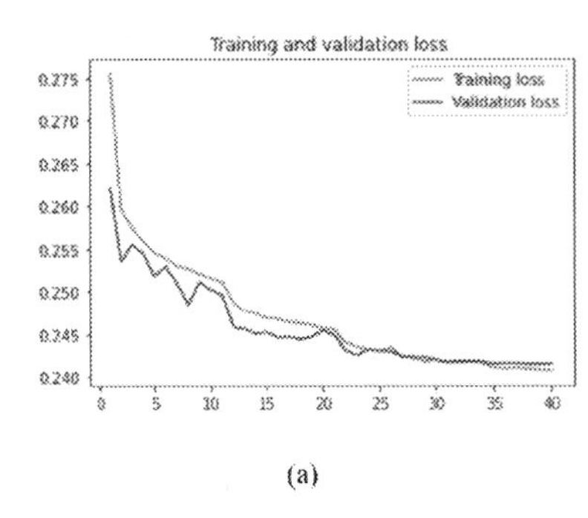

(a)

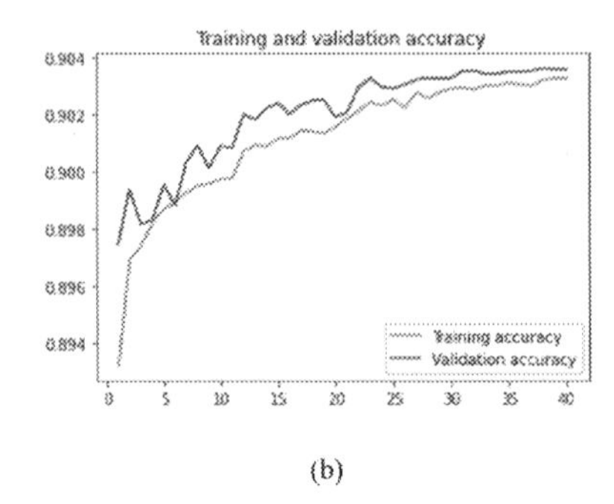

(b)

Figure 12.9 Loss and accuracy curves of the Xception architecture in (a) and (b) respectively.

12.4.3 Results and Discussion

In using the transfer learning method, we considered the pre-trained weights obtained from the IMAGENET dataset. In our proposed architecture, we have chosen the hard sharing of features method. The loss and accuracy curves are shown in Figure 12.9.

As for multi-label classification, a single confusion matrix cannot describe the result. So for every label, a confusion matrix is required. Keeping this in mind, we show the confusion matrix for our proposed method in Figure 12.10. A Convolutional Neural Network's hyperparameters are those that the model can't discover on its own. After many trials and errors, we came up with the following hyperparameters for our proposed architecture. Different values of these hyperparameters caused higher loss and lower accuracy. The reason for these errors is the quick convergence of the training data. So with experience and a lot of trials, these hyperparameters are chosen for our proposed work. The initial learning rate is 0.01 with factor 0.5, patience 3, minimum learning rate $1e^{-8}$, and minimum delta $1e^{-4}$. A dense layer with 1024 nodes is used in our proposed work. A batch size of 64 and 40 epochs are used in the experiment. The Adam optimizer is used as it gives better performance than other optimizers in most cases. The hyperparameters are selected from the corresponding hyperparameter space as shown in Table 12.3. These selected values for the hyperparameters gave the best accuracy, i.e., 92.52% accuracy, and the lowest loss value. Binary cross-entropy loss is used here as this is a binary classification task.

The ROC curve in Figure 12.11 shows a graphical representation of the True positive rate and False positive rate for the pedestrian attribute recognition task.

For the proposed method, we selected 54 binary attributes. The reason behind this is that previous works are also done with these 54 attributes, so we can compare with their work because we included the same attributes. Table 12.4 shows the comparison with SOTA (state-of-the-art) methods. We can see from Table 12.4 that our proposed method performs better than other methods. In our proposed architecture, we have chosen the hard sharing of features method. Hard sharing means sharing the features for all our selected attributes in the Fully Connected layers. This technique considers the inter-dependency among attributes. This is one of the reasons for the better result. Also, the feature extraction process of a Convolutional Neural Network can be

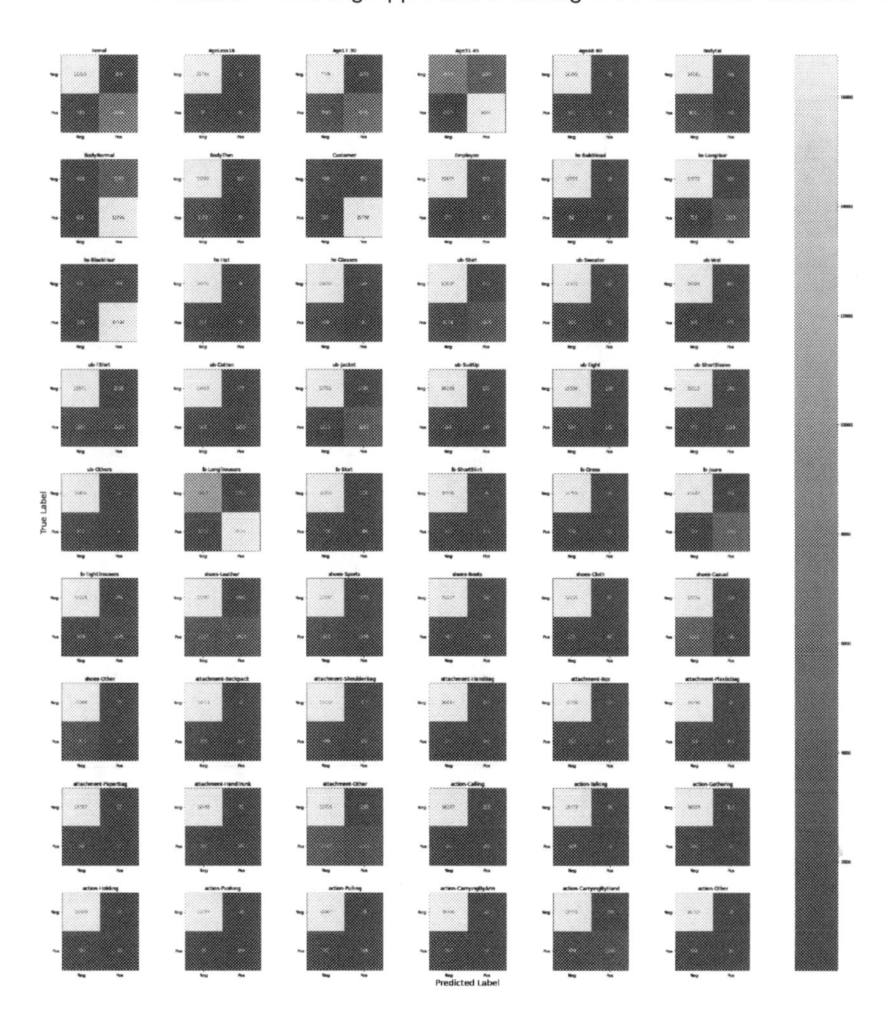

Figure 12.10 Confusion matrix of the Xception architecture.

Table 12.3 Optimum hyperparameter selection from hyperparameter space

Hyperparameters	Hyperparameter Space	Optimum Hyperparameter
Number of Dense Layer	1, 2	1
Neurons in Dense Layer	256, 512, 1024, 2048	1024
Learning Rate	0.1, 0.01, 0.001, 0.5, 0.05, 0.005	0.01
Batch Size	32, 64, 128	64
Epochs	20, 30, 40, 50	40
Optimizer	'Adam', 'RMSProp', 'SGD'	'Adam'

compared to a black box as described in [23]. Because it is a black box, the information is dispersed in a complicated way to comprehend. For these reasons, the overall accuracy is 92.52%, which is better than the SOTA methods. The demonstration of the proposed method is shown in Figure 12.12. The source of the image is in [11].

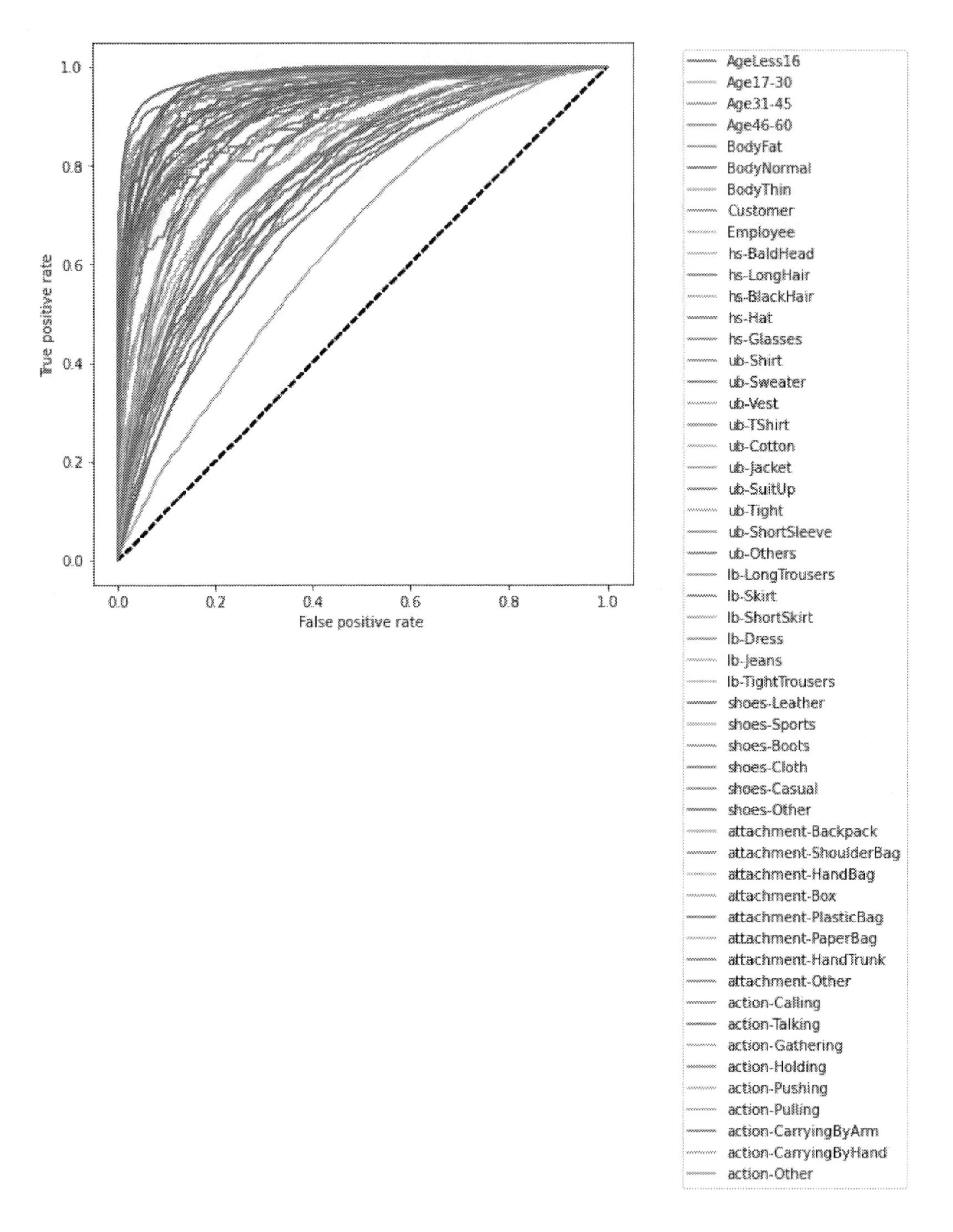

Figure 12.11 ROC curves for pedestrian attribute recognition.

In short, Table 12.1 shows that among other pre-trained architectures, our proposed Xception architecture performs better. After that, experiments performed on the proposed architecture increased the accuracy in Table 12.2. Table 12.3 illustrates the optimum hyperparameters. The hyperparameters are one dense layer, 1024 nodes

Table 12.4 Comparison with state-of-the-art methods on the RAP v2 dataset.

Method	RAP V2 Dataset
	mA
[11]	68.92%
[10]	75.54%
[21]	77.87%
[22]	76.74%
Proposed method	**92.52%**

Figure 12.12 Demonstration of our proposed method: (a) input image, (b) output image.

in the dense layer, 0.01 learning rate, 64 batch size, 40 epochs, and the Adam optimizer. Finally, a comparison with the SOTA methods in Table 12.4 shows that our proposed method can produce better results than other methods by applying the transfer learning technique with a robust CNN architecture. Also, fine-tuning the hyperparameters can increase the accuracy. So our hypothesis for proposing the method is justified. However, as the CNN architecture is powerful, the network parameter is large. It takes a good amount of time to recognize the pedestrian attributes.

12.5 CONCLUSION

In our method, we have proposed a pre-trained CNN architecture, Xception, with a transfer learning approach. We fine-tuned the hyper-parameters to obtain better results. Moreover, the positive label distribution of the RAP v2 dataset is also shown in our proposed method. The loss and accuracy curve and the confusion matrix for each attribute are shown too. Our proposed CNN architecture outperformed existing SOTA methods with 92.52% accuracy. However, the proposed CNN architecture has large network parameters, which is approximately 8.9M. This large number of parameters increases the computation cost for computers. Because the RAP v2 dataset is imbalanced, we will likely work with a more balanced dataset in the future. Also, more hyperparameter tuning might improve the results. Additionally, we intend to extend our work to multiple pedestrian scenarios.

Bibliography

[1] Wang X, Zheng S, Yang R, et al. Pedestrian attribute recognition: A survey. arXiv preprint arXiv:190107474. 2019;.

[2] Tian Y, Luo P, Wang X, et al. Pedestrian detection aided by deep learning semantic tasks. In: *Proceedings of the IEEE Conference on Computer Vision and Pattern Recognition*; 2015. p. 5079–5087.

[3] Lin Y, Zheng L, Zheng Z, et al. Improving person re-identification by attribute and identity learning. *Pattern Recognition*. 2019;95:151–161.

[4] Khan A, Sohail A, Zahoora U, et al. A survey of the recent architectures of deep convolutional neural networks. *Artificial Intelligence Review*. 2020;53(8):5455–5516.

[5] Fang W, Chen J, Hu R. Pedestrian attributes recognition in surveillance scenarios with hierarchical multi-task CNN models. *China Communications*. 2018;15(12):208–219.

[6] Gray D, Tao H. Viewpoint invariant pedestrian recognition with an ensemble of localized features. In: *European Conference on Computer Vision*; Springer; 2008. p. 262–275.

[7] Prosser BJ, Zheng WS, Gong S, et al. Person re-identification by support vector ranking. In: *BMVC*; Vol. 2; 2010. p. 6.

[8] Layne R, Hospedales TM, Gong S, et al. Person re-identification by attributes. In: *BMVC*; Vol. 2; 2012. p. 8.

[9] Layne R, Hospedales TM, Gong S. Attributes-based re-identification. In: *Person Re-identification*. Springer; 2014. p. 93–117.

[10] Li D, Chen X, Huang K. Multi-attribute learning for pedestrian attribute recognition in surveillance scenarios. In: *2015 3rd IAPR Asian Conference on Pattern Recognition* (ACPR); IEEE; 2015. p. 111–115.

[11] Sudowe P, Spitzer H, Leibe B. Person attribute recognition with a jointly-trained holistic CNN model. In: *Proceedings of the IEEE International Conference on Computer Vision Workshops*; 2015. p. 87–95.

[12] Kurnianggoro L, Jo KH. Identification of pedestrian attributes using deep network. In: *IECON 2017-43rd Annual Conference of the IEEE Industrial Electronics Society*; IEEE; 2017. p. 8503–8507.

[13] Yu W, Kim S, Chen F, et al. Pedestrian detection based on improved mask R-CNN algorithm. In: *International Conference on Intelligent and Fuzzy Systems*; Springer; 2020. p. 1515–1522.

[14] Cao X, Wang Z, Yan P, et al. Transfer learning for pedestrian detection. *Neurocomputing*. 2013;100:51–57.

[15] He K, Gkioxari G, Dollár P, et al. Mask R-CNN. In: *Proceedings of the IEEE International Conference on Computer Vision*; 2017. p. 2961–2969.

[16] Redmon J, Divvala S, Girshick R, et al. You only look once: Unified, real-time object detection. In: *Proceedings of the IEEE Conference on Computer Vision and Pattern Recognition*; 2016. p. 779–788.

[17] Bharati P, Pramanik A. Deep learning techniques—R-CNN to mask R-CNN: A survey. In: *Computational Intelligence in Pattern Recognition*. Springer; 2020. p. 657–668.

[18] Team K. Keras documentation: Keras applications ; 05/13/2021. https://keras.io/api/applications/.

[19] Li D, Zhang Z, Chen X, et al. A richly annotated dataset for pedestrian attribute recognition. arXiv preprint arXiv:160307054. 2016.

[20] Li D, Zhang Z, Chen X, et al. A richly annotated pedestrian dataset for person retrieval in real surveillance scenarios. *IEEE Transactions on Image Processing*. 2018;28(4):1575–1590.

[21] Sarafianos N, Xu X, Kakadiaris IA. Deep imbalanced attribute classification using visual attention aggregation. In: *Proceedings of the European Conference on Computer Vision* (ECCV); 2018. p. 680–697.

[22] Guo H, Zheng K, Fan X, et al. Visual attention consistency under image transforms for multi-label image classification. In: *Proceedings of the IEEE/CVF Conference on Computer Vision and Pattern Recognition*; 2019. p. 729–739.

[23] Can we open the black box of AI? 07/05/2021. https://www.nature.com/news/can-we-open-the-black-box-of-ai-1.20731.

CHAPTER 13

TF-IDF Feature-based Spam Filtering of Mobile SMS Using a Machine Learning Approach

Syed Md. Minhaz Hossain

Department of Computer Science & Engineering, Premier University, Chattogram, Bangladesh

Khaleque Md. Aashiq Kamal

Department of Computer Science & Engineering, Premier University, Chattogram, Bangladesh

Anik Sen

Department of Computer Science & Engineering, Premier University, Chattogram, Bangladesh

Iqbal H. Sarker

Department of Computer Science & Engineering, Chittagong University of Engineering & Technology, Chattogram, Bangladesh

CONTENTS

SHort Message Service (SMS) is becoming the secure medium of communication due to large-scale global coverage, reliability, and power efficiency. As person-to-person (P2P) messaging is less secure than application-to-person (A2P) messaging, anyone can send a message, leading an attack. Attackers mistreat this opportunity by spreading malicious content, performing harmful activities, and abusing other people, which is commonly known as spam. Moreover, such messages can waste a lot of time, and important messages are sometimes overlooked. As a result, accurate spam detection in SMS and its computational time are burning issues. In this chapter, we conduct six different experiments to detect SMS spam from a dataset of 5574 messages using machine learning classifiers such as Multinomial Naive Bayes (MNB) and Support Vector Machine (SVM), and consider variations of Term Frequency-Inverse Document Frequency (TF-IDF) features for exploring the trade-offs among accuracy, F1-score and computational time. The experiments achieve the best accuracy of 98.50%, F1-score of 98%, and area under the ROC curve (AUC) of 0.97 for a multinomial naive Bayes classifier with TF-IDF after stemming.

13.1 INTRODUCTION

The rise of 5G and cloud technology introduces a new ecosystem that incorporates the connectivity of devices and technologies. SMS is becoming the secure medium for machine-to-machine (M2M) or machine-to-person (M2P) communication for large-scale global coverage, reliability, power efficiency, and reduction of SMS cost [2]. In addition, SMS has a vital role in making a decision based on the data via the Internet of Things (IoT) [5]. IoT users can get notifications and other alarms from IoT devices through SMS [6]. As the smart devices used in our everyday life activities are mostly directed by internet connectivity, the risk of data privacy or cyber-attacks is increasing day by day [16]. Cyber-attacks cause vital infrastructural destruction with massive losses of $345 per incident [1]. Spamming is an effective way to spread malware through the internet and mobile networks. Usually, a mobile user faces a crisis of spamming through SMS. This fact has led attackers to use SMS spam to spread cyber-attack payloads [3]. Moreover, SMS spam is considered the most straightforward technique for deploying phishing attacks [4]. As a result, security specialists are very devoted to developing an efficient SMS spam detection or classification method.

Machine learning techniques such as SVM, decision tree (DT), logistic regression (LR), and MNB play an essential role in detecting anomalies or classifying SMS spam [15, 17, 18, 19, 24]. Moreover, these techniques have made vital contributions to detecting email spam messages [21, 20]. Several studies have been done on the different types of proposed methods for filtering mobile SMS spam [25, 26, 27, 28]. Different feature extraction methods such as word2vec, word n-gram, character n-gram, and combinations of variable length n-grams, are used to extract features in several works [22, 23, 31]. Moreover, in spite of having the highest accuracy in machine learning- based spam SMS detection [29, 30], the ratio of false positives (precision) to false negatives (recall) is an issue.

Moreover, in Gupta et al. (2018), a deep learning algorithm is used to detect spam SMS and achieves 99.1% accuracy [32]. Using deep learning in detecting spam SMS is computationally expensive and time-consuming, requires a large amount of data. Despite the better accuracy of the deep learning method, it is a burning question of high accuracy in less time. Machine learning techniques consume less time than deep learning-based spam SMS detection. Bagging and boosting algorithms achieve better results than traditional machine learning methods [31]. However, there are still limitations of computational complexity and loss of interpretability. Moreover, TF-IDF has problems finding similar words for each document. Despite this limitation, TF-IDF is significant for its improvement in the ratio of false positives (precision) to false negatives (recall) [31] in its formulation.

As a result, for exploring the trade-offs among accuracy, F1-score, and time, we conducted six different experiments to detect SMS spam from the dataset using machine learning classifiers such as MNB and SVM. The primary contributions of this chapter are as follows:

(i) Develop a spam filter for SMS using trade-offs among accuracy, F1-score, and computational time.

(ii) Investigate a different feature extraction process using TF-IDF and evaluate the impact of stemming before and after determining TF-IDF scores.

(iii) Investigate the effect of message length in feature extraction.

(iv) Determine the best classification model with the most appropriate feature extraction process.

The other sections of this chapter have been arranged in the following manner: in Sect. 13.2, the previous benchmark works are discussed. The proposed method for detecting spam is presented in Sect. 13.3. In Sect. 13.4, experiments, results, and observations are illustrated. Finally, the chapter is concluded in Sect. 13.5.

13.2 RELATED WORK

In this section, we will look at comparative studies that other researchers have conducted. Joe et al. [7] developed a mobile SMS spam filtering system. They used 460 messages as a dataset, with 300 messages to train the SVM model and 160 for testing. Their model produced the best results with a feature vector value of 150 and a gamma value of 0.01 gamma. In [8], researchers worked on 875 messages as a data source, consisting of Turkish and English languages. They showed that BoW and Sfs together give the F1-score effectively. KNN and SVM are used as classifier models in their experiments. On the other hand, Foozy et al. [9] used naive Bayes and J48 as classifier models to conduct their experiments. They selected the Malay language to generate their data set. Both naive Bayes and J48 models have given promising results. The text processing approach has been deployed in the work of [10]. They used a public dataset with 5,574 the English language SMSs. Different classification models, including B.C4.5, KNN, SVM, naive Bayes, and PART, have been used in this work.

Moreover, Reaves et al. [11] used only the SVM as a classifier model on a labeled SMS dataset that was gathered over 400 days. Their proposed classification model achieved 96.8% recall and was highly precise. In [12], researchers have used a probabilistic data structure (PDS) for ensemble-based spam classification. They also showed that the number of spam messages could hamper social networks in the field of IoT. Arifin et al. [13] worked on two different datasets consisting of 5,574 and 1,324 messages. Naive Bayes classifier model is used in their work. To improve performance, they used FP-Growth as an association rule. This collaboration has given the highest accuracy value, with 98%. In [29], the authors used TF−IDF for feature extraction and random forest as classifiers. This work achieves 97.50% accuracy. In [30], machine learning classifiers such as logistic regression (LR), K-nearest neighbor (K-NN), and decision tree (DT) are used for spam classification in mobile device communication. The performance of LR achieved a high accuracy of 99%.

13.3 MATERIALS AND METHODS

At first, a dataset of spam and ham was collected. Then, we applied the pre-processing and feature extraction method to the dataset. The extracted features have been used to create a feature vector for training and testing the classifier model. The classifier model determines whether an SMS is spam or ham. The general structure of our model is shown in Figure 13.1.

13.3.1 Pre-processing

Pre-processing is one of the significant phases in text classification. Pre-processing eliminates noises and unwanted words from raw text and structures the text. Sometimes, unwanted words and features can affect the performance of various probabilistic

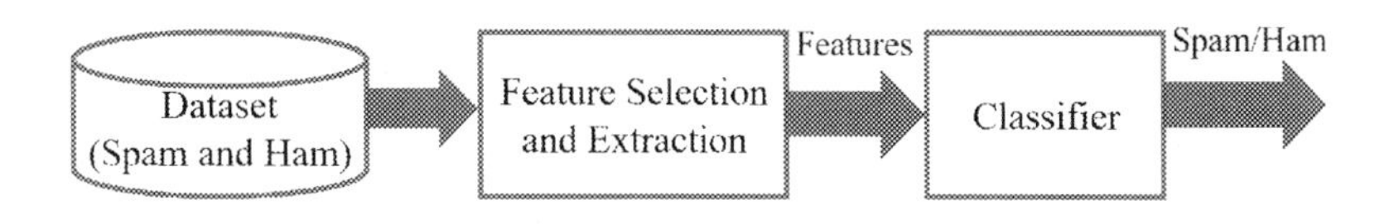

Figure 13.1 The overall procedure of the TF-IDF.

classifiers. In this section, we depict all the pre-processing steps done in our work with an example.

Consider an SMS: ("hi! we bring a software....")

13.3.1.1 Redundant character removal

For each SMS, special characters such as (,=,>), numbers, and punctuation are removed. So, the SMS might become ("hi we bring a software").

13.3.1.2 Removal of stop words

Then from each SMS, we remove various words like "the," "an," "this," "a," and others that do not have meaning in extracting the features of spam or ham. Finally, the output would be ("bring software").

13.3.1.3 Tokenization

In this phase, we break a large block of text into a list of words such as ("bring", "software").

13.3.1.4 Lemmatization

Lemmatization eliminates inflectional morphemes and stores the lemma. A lemma is basically a base or a dictionary form of a phrase or word. After this phase, it would be ("bring", "software").

13.3.2 Feature Extraction

Our method explores features based on three features: (1) Length of the message, (2) TF-IDF, (3) Stemming.

1. Length of the messages: The length of the message is a crucial feature to determine the nature of any message. In this extraction phase, the length of all messages has been calculated using the number of remaining words after the pre-processing step.

2. TF-IDF: TF-IDF is another vital extraction process to evaluate the nature of any SMS. Pre-processing is performed before starting the process of the TF-IDF algorithm.

    i) After the pre-processing step, tokenization is performed on sentences instead of words. Then the weight value is assigned.

    ii) Then, the calculation of word frequency is done.

    iii) In this step, TF (Term Frequency) is calculated using the formula in Eq. 13.1.

$$TF(x) = \frac{(\text{No. of times words } x \text{ in a doc})}{(\text{Total no. of words in that doc})} \tag{13.1}$$

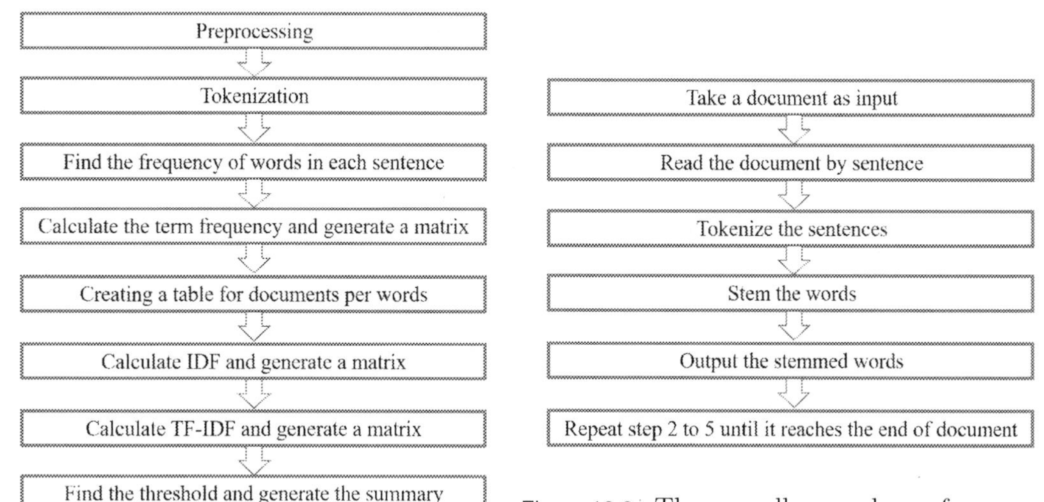

Figure 13.2 The overall procedure of the TF–IDF process.

Figure 13.3 The overall procedure of Stemming process.

iv) After that, a table is created for the frequency of every word in every sentence.

v) Then, the IDF (Inverse document frequency) is calculated using Eq. 13.2.

$$IDF(x) = \log_e \frac{\text{Total no. of docs}}{\text{No. of docs with } x \text{ in it}} \tag{13.2}$$

vi) TF-IDF is calculated by multiplying the value of Eq. 13.1 and Eq. 13.2.

vii) In these steps, the average score of all words is calculated to find the threshold value. Finally, any words are selected if the corresponding score is higher than the threshold value. Figure 13.2 concludes the overall procedure of the TF-IDF process.

3. Stemming: Finally, we use the stemming process before extracting TF-IDF scores. It is a way to find the root of any word. By eliminating prefixes and suffixes, we can get the root of a word. Figure 13.3 concludes the overall procedure of the stemming process.

13.3.3 Classifiers

After selecting three different feature extraction processes, we conducted six individual experiments using two classifier models: the support vector machine (SVM) and multinomial naive Bayes (MNB).

13.3.3.1 Support Vector Machine

The Support vector machine is a supervised machine learning model used in various machine learning applications due to its higher accuracy. The SVM classifies data in a binary classification problem by determining the best hyperplane that separates all data points of one class from those of the other. The goal of the SVM is to maximize the separated hyperplane using a formula as shown in Eq. 13.3 to differentiate the classes:

$$w.x_i + b = 0 \tag{13.3}$$

where w is the weight factor, b is the bias, and x is the feature vector of sample i.

13.3.3.2 Multinomial Naive Bayes

The Naive Bayes algorithm is computationally efficient when used in text analysis. Naive Bayes gives all words equal importance to track every single phrase in text. Naive Bayes uses Bayes' theorem, where features are mutually independent. That means the probability of one feature does not depend on the other feature. The probability model is formulated in Eq. 13.4.

$$P(A|B) = P(A) \times P(B|A)/P(A) \tag{13.4}$$

where P(B) is the prior probability of B, P(A) = prior probability of A and P(B|A) = occurrence of predictor B given class A probability.

13.4 RESULT AND OBSERVATIONS

All of the experiments were carried out on a computer with an Intel Core i5 processor and 4 GB of RAM. We have implemented the experiments using Python 3.7 on the Spider IDE.

13.4.1 Dataset

We used an open dataset of SMS messages from the kaggle [14]. The dataset is labeled with two classes: spam and ham. It contains a total of 5,572 messages, of which 4,825 are ham and 747 are spam (4516 unique ham messages and 653 spam messages). We divide the dataset into 70-30, 67-33, 80-20 proportion for training purposes. As the 70-30 proportions attain better accuracy, we consider 70-30% of training and test messages by different classifiers.

13.4.2 Classification Using SVM and Multinomial Naive Bayes

The calculation of the TF-IDF score and the TF-IDF after stemming for a sample SMS (labeled as ham) from our examined dataset are shown in Figure 13.4 and Figure 13.5. SVM and MNB classifiers are used to incorporate all three features of messages. At first, to select the best hyper-parameters, a grid search method is used. In SVM, the best Gamma is defined as 1 with the sigmoid kernel. The value of Alpha

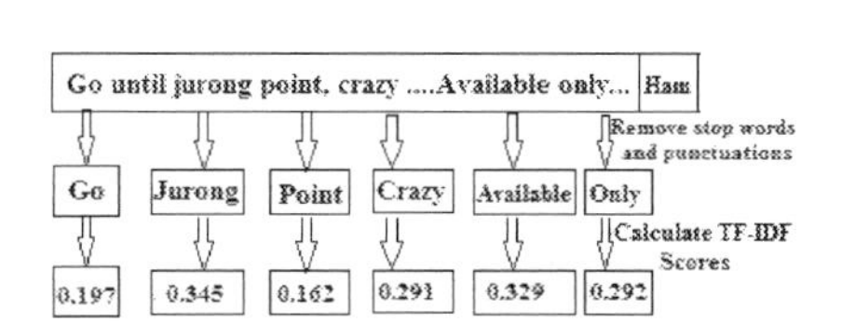

Figure 13.4 An example for TF-IDF score calculation.

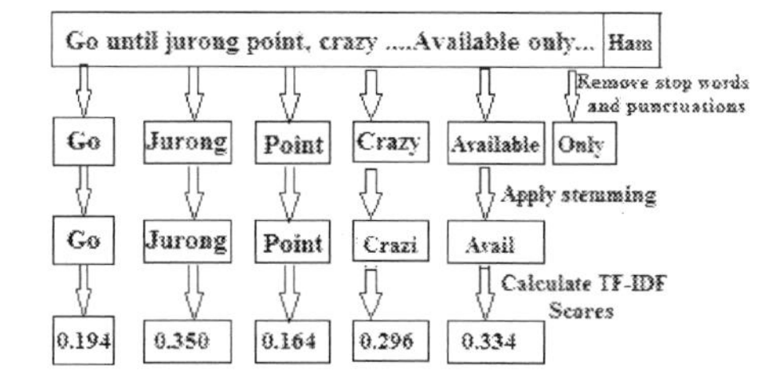

Figure 13.5 An example for TF-IDF score calculation after applying the stemming process.

is set to 0.2 in MNB. Six different experiments are debated with critical analysis based on these hyper-parameters.

13.4.2.1 Performance Measure

As our ham and spam samples are imbalanced, we consider accuracy and the F1-score for evaluating our detection model as formulated in Eq. (13.5)-(13.11).

$$\text{Accuracy of a model} = \frac{\sum_k(\text{Recognition rate of each class} \times N_k)}{N} \tag{13.5}$$

$$\text{Recognition rate of a class} = \frac{\text{True Positive} + \text{True Negative}}{\text{Number of all samples of that class}} \tag{13.6}$$

$$\text{Precision of a model} = \frac{\sum_k(\text{ Precision of each class} \times N_k)}{N} \tag{13.7}$$

$$\text{Precision of a class} = \frac{\text{True Positive}}{\text{True Positive} + \text{False Positive}} \tag{13.8}$$

$$\text{Recall of a model} = \frac{\sum_k(\text{Recall of each class} \times N_k)}{\text{N}} \tag{13.9}$$

$$\text{Recall of a class} = \frac{\text{True Positive}}{\text{True Positive} + \text{False Negative}} \tag{13.10}$$

$$F_1\text{-score} = 2 \times \frac{\text{Precision} \times \text{Recall}}{\text{Precision} + \text{Recall}} \tag{13.11}$$

where k represents each class, N_k indicates the number of samples in class k, and N is the total number of samples used to test the model.

13.4.2.2 Performance Evaluation for Different Feature Extraction Methods Using Various Classifiers

The accuracy and F1-score of six different experiments are summarized in Table 13.1. With only the TF-IDF score as feature extraction, both SVM and MNB achieve remarkable accuracy of 97.85% and 98.45%, respectively. The SVM and MNB models are 97.85% and 98.50% accurate for the TF-IDF score with stemming, respectively.

Table 13.1 Performance evaluations of all the experiments.

Exp.	Feature Extraction	Classifier	Accuracy	F1-score	Computational Time
1.	TF-IDF	SVM	97.85%	98%	0.545 sec
2.	TF-IDF with stemming	SVM	97.85%	98%	0.484 sec
3.	Stemming, TF-IDF and Length of messages	SVM	86.12%	80%	0.567 sec
4.	TF-IDF	MNB	98.45%	98%	0.021 sec
5.	TF-IDF with stemming	MNB	98.50%	98%	0.023 sec
6.	Stemming, TF-IDF and Length of messages	MNB	98.27%	98%	0.128 sec

Table 13.2 Performances of each class in detecting spam messages using MNB with TF-IDF features(extracted after stemming).

Classes	**Precision**	**Recall**	**F1-score**
Ham	100%	**96%**	98%
Spam	96%	**100%**	98%

In this case, we can say that using the TF-IDF score with stemming improves the accuracy of the MNB classifier model by 0.05%. Finally, to compare the results, we add another feature (message length). However, in this case, the SVM model's accuracy has dropped to 86.12%, whereas MNB's accuracy has remained consistent at 98.27%. As a result, we can deduce that factoring message length into the experiment adds no value.

However,the SVM with stemming, TF-IDF scores, and length of the messages give the lowest F1-score, which is 80%. The other two cases give a better F1-score with a value of 98%. F1-scores are also consistent for the MNB classifier model for all three cases with a value of greater than or equal to 98%.

From Table 13.1, it is obvious that with stemming, TF-IDF scores are better for both SVM and MNB. MNB achieves better performance for TF-IDF integrated with stemming as stemming traces the different terms of the same meaning significantly.

13.4.2.3 Performance Representation for the Best Classifier Using AUC and Confusion Matrix

We test 1115 messages, including 1003 ham messages and 112 spam messages using both the classifiers SVM and MNB with variations of TF–IDF. As we achieve the best accuracy using MNB with TF–IDF features (extracted after stemming), it is shown in Table 13.2 for individual classes. MNB achieves better accuracy in every case than SVM due to the performance loss for imbalances in positive and negative support vectors in SVM.

To evaluate the results further, we have drawn the ROC curves for both cases. The micro average area under the ROC curve (AUC) for SVM is 0.93, and for the MNB is 0.97. Figure 13.6 shows the area under the ROC curve (AUC) for the MNB

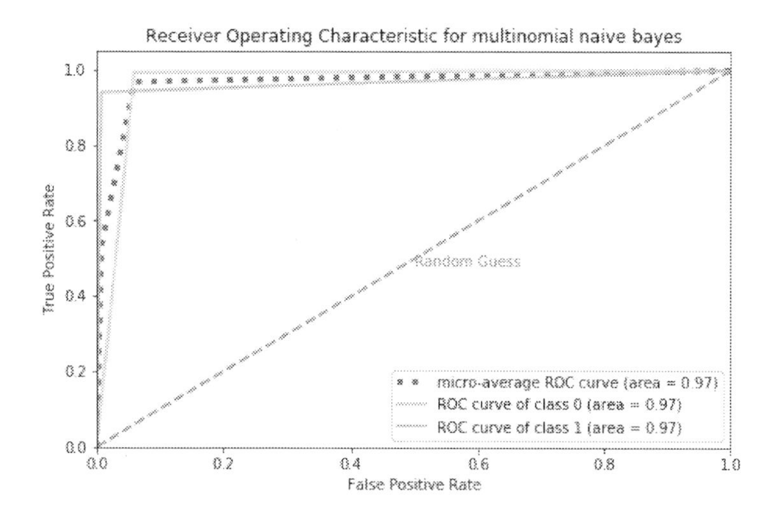

Figure 13.6 ROC curve for multinomial naive Bayes classifier model using TF-IDF integrated with stemming.

Table 13.3 Comparison of our work with other benchmark works.

Reference	Feature Extraction	Classifier	Accuracy	F1-score
[29]	TF-IDF	Random Forest	94.5%	96.24%
[30]	TF-IDF	Logistic Regression	96.8%	97.00%
Our method	TF-IDF	Multinomial Naive Bayes	98.50%	98.00%

classifier. It implies that MNB is more robust for filtering the spam in SMS with stemming and the TF-IDF process.

13.4.2.4 Computational Time Analysis for Classifying Spam

The impact of feature selection on computational time is reflected in Table 13.1. The computational time for SVM drops from 0.5445 seconds to 0.484 seconds after stemming. In MNB, however, it is almost the same as stemming. However, considering all the factors (accuracy, F1-score, and computational time), MNB is the optimal model for detecting spam SMS using TF-IDF integrated with stemming.

13.4.2.5 Comparison Among the Benchmark Spam Detection Methods

We compare our work with the other benchmark methods, as shown in Table 13.3. We implement their feature extraction technique and classification models on our dataset to compare with other benchmark works. Our proposed method achieves better accuracy of 98.50% and F1-score of 98% than the other two works.

13.4.2.6 Critical Evaluation

Despite the fact that our proposed method provides better accuracy and F1-score, it still has some misclassifications. Consider the following example of a misclassified ham message: "Ok lar... Joking wif u oni." The majority of the words in this message are removed during the pre-processing stage. In the feature extraction process, only "Joke" is counted. Despite the fact that it is a ham message, it is flagged as spam. In this case, the message's context is critical. We will also use context-based machine learning to look into the features of the messages.

13.5 CONCLUSION

For security in message communication, detection of spam plays an important role. Accuracy and efficiency in detecting spam are significant issues to allow users to read only relevant messages. However, to solve these issues, we investigate all the possibilities of improving SMS classification accuracy using variations of TF-IDF. Best scores of accuracy (98.50%) and F1-score (98%) have been found by incorporating stemming and TF–IDF using MNB. In addition, we conclude that message length does not matter in spam detection in SMS. When we use stemming, it also increases the accuracy in MNB; however, execution time remains constant for filtering SMS. Finally, we can conclude that MNB shows better results than SVM in all cases, whether the size of words decreases or the length of messages increases or decreases. We will try to extend our work by detecting text embedded in an image (stego image) and developing a method using ontology and the semantic web for scalable SMS spam filtering.

Bibliography

[1] Hu, X.: Large-scale malware analysis, detection, and signature generation. [online] Available at: https://https://deepblue.lib.umich.edu/handle/2027.42/89760 [Accessed 28 Feb. 2020]

[2] Lau, C. Q., Cronberg, A., Marks, L., & Amaya, A. (2019). In Search of the Optimal Mode for Mobile Phone Surveys in Developing Countries. A Comparison of IVR, SMS, and CATI in Nigeria. *Survey Research Methods*, 13(3), 305–318.

[3] Dang. That wasn't supposed to happen. [online] Available at: https://mobilecommons.com/blog/2016/01/how-text-messaging-will-change-for-the-better-in-2016 [Accessed 28 Feb. 2020]

[4] Gupta, B. B., Aakanksha Tewari, Ankit Kumar Jain, and Dharma P. Agrawal. (2016). Fighting against phishing attacks: State of the art and future challenges. *Neural Computing & Applications*, 28 (12): 3629–54. https://doi.org/10.1007/s00521-016-2275-y.

[5] Atzori, L., Iera, A., Morabito, G., Nitti, M.(2012). The Social Internet of Things (SIoT) – When social networks meet the Internet of Things: Concept, architecture and network characterization. *Computer Networks*, 56, 3594–3608.

[6] M&M Research Group, 2012. Internet of Things (IoT) & M2M communication market-advanced technologies, future cities & adoption trends, roadmaps & worldwide forecasts 2012- 2017. [online] Available at: https://www.prnewswire.com/news-releases/internet-of-things-iot–machine-to-machine-m2m-communication-market—advanced-technologies-future-cities–adoption-trends-roadmaps–worldwide-forecasts-2012—2017-216448061.html [Accessed 28 Feb. 2020]

[7] Joe, I., Shim, H.(2010). An SMS spam filtering system using support vector machine. *Future Generation Information Technology*, 577–584.

[8] Uysal, A.K., Gunal, S., Ergin, S., Sora Gunal, E.(2013). The impact of feature extraction and selection on SMS spam filtering. *Electronics and Electrical Engineering*, 19.

[9] Mohd Foozy, C.F., Ahmad, R. and Abdollah, M.F.(2014). A framework for SMS spam and phishing detection in the Malay language: A case study. *International Review on Computers and Software*, 9(7), pp. 1248–1254.

[10] Almeida, T.A., Silva, T.P., Santos, I., Gómez Hidalgo, J.M.(2016). Text normalization and semantic indexing to enhance Instant Messaging and SMS spam filtering. *Knowledge-Based Systems*, 108, 25–32.

[11] Reaves, B., Blue, L., Tian, D., Traynor, P., Butler, K.R.B.(2016). Detecting SMS spam in the age of legitimate bulk messaging. *Proceedings of the 9th ACM Conference on Security & Privacy in Wireless and Mobile Networks.*

[12] Singh, A., Batra, S.(2018). Ensemble based spam detection in social IoT using probabilistic data structures. *Future Generation Computer Systems*, 81, 359–371.

[13] Delvia Arifin, D., Shaufiah, Bijaksana, M.A.(2016). Enhancing spam detection on mobile phone Short Message Service (SMS) performance using FP-growth and naive Bayes classifier. *2016 IEEE Asia Pacific Conference on Wireless and Mobile (APWiMob).*

[14] SMS Spam Collection Dataset. [online] Available at: https://www.kaggle.com/uciml/sms-spam-collection- dataset [Accessed 28 Feb. 2020]

[15] Sarker, Iqbal H. (2021). Machine learning: Algorithms, real-world applications and research directions. *SN Computer Science*, 2.3, 1–21.

[16] Sarker, Iqbal H., Md Hasan Furhad, and Raza Nowrozy.(2021). AI-driven cybersecurity: An overview, security intelligence modeling and research directions. *SN Computer Science*, 2.3, 1–18.

[17] Sarker, Iqbal H. (2021). CyberLearning: Effectiveness analysis of machine learning security modeling to detect cyber-anomalies and multi-attacks. *Internet of Things*, 14, 100393.

[18] Dada, E.G.; Bassi, J.S.; Chiroma, H.; Adetunmbi, A.O.; Ajibuwa, O.E. (2019). Machine learning for email spam filtering: Review, approaches and open research problems. *Heliyon*, 5, e01802.

[19] Shah, N.F.; Kumar, P. (2018). A comparative analysis of various spam classifications. *In Progress in Intelligent Computing Techniques: Theory, Practice, and Applications; Springer: Berlin/Heidelberg, Germany*, pp. 265–271.

[20] Chandrasekar, C.; Priyatharsini, P. (2018). Classification techniques using spam filtering email. *Int. J. Adv. Res. Comput. Sci.*, 9, 402.

[21] Shafi'I, M.A.; Latiff, M.S.A.; Chiroma, H.; Osho, O.; Abdul-Salaam, G.; Abubakar, A.I.; Herawan, T. (2017). A review on mobile SMS spam filtering techniques. *IEEE Access*, 5, 15650–15666.

[22] J. Hua and Z. Huaxiang (Mar.2015). Analysis on the content features and their correlation of Web pages for spam detection, *China Commun.*, vol. 12, no. 3, pp. 84–94.

[23] S.-E. Kim, J.-T. Jo and S.-H. Choi (2015). SMS spam filtering using keyword frequency ratio, *Int. J. Secur. Appl.*, vol. 9, no. 1, pp. 329–336.

[24] E. B. Cleff (2007). Privacy issues in mobile advertising. *Int. Rev. Law Comput. Technol.*, vol. 21, pp. 225–236.

[25] O. Osho, O. Y. Ogunleke and A. A. Falaye (Oct. 2014). Frameworks for mitigating identity theft and spamming through bulk messaging. *Proc. IEEE 6th Int. Conf. Adapt. Sci. Technol. (ICAST)*, pp. 1–6.

[26] R. Islam and J. Abawajy (Jan. 2013). A multi-tier phishing detection and filtering approach. *J. Netw. Comput. Appl.*, vol. 36, no. 1, pp. 324–335.

[27] O. Osho, V. L. Yisa, O. Y. Ogunleke and S. I. M. Abdulhamid (Nov. 2015). Mobile spamming in Nigeria: An empirical survey. *Proc. Int. Conf. Cybersp. (CYBER-Abuja)*, pp. 150–159.

[28] Sarker, Iqbal H.(2021). Data Science and Analytics: An overview from data-driven smart computing, decision-making and applications perspective. *SN Computer Science*, 2, 377.

[29] Nilam Nur Amir Sjarif, Nurulhuda Firdaus Mohd Azmi, Suriayati Chuprat, Haslina Md Sarkan, Yazriwati Yahya, Suriani Mohd Sam (2019). SMS spam message detection using term frequency-inverse document frequency and random forest algorithm. *The Fifth Information Systems International Conference*, 509–515.

[30] Luo GuangJun, Shah Nazir, Habib Ullah Khan, and Amin Ul Haq (2020). Spam detection approach for secure mobile message communication using machine learning algorithms. *Security and Communication Networks* , vol. 2020, Article ID 8873639, 6 pages, https://doi.org/10.1155/2020/8873639.

[31] Kowsari, Kamran, Kiana Jafari Meimandi, Mojtaba Heidarysafa, Sanjana Mendu, Laura Barnes, and Donald Brown (2019). Text classification algorithms: A survey. *Information*, 10, no. 4, 150.

[32] Gupta, M., Bakliwal, A., Agarwal, S. and Mehndiratta, P. (2018). A comparative study of spam SMS detection using machine learning classifiers. *11th International Conference on Contemporary Computing*, IC3 2018 pp. 1–7.

CHAPTER 14

Content-based Spam Email Detection Using an N-gram Machine Learning Approach

Nusrat Jahan Euna

Department of Computer Science & Engineering, Chittagong University of Engineering & Technology, Chattogram, Bangladesh

Syed Md. Minhaz Hossain

Department of Computer Science & Engineering, Premier University, Chattogram, Bangladesh

Md. Musfique Anwar

Department of Computer Science and Engineering, Jahangirnagar University, Dhaka, Bangladesh

Iqbal H. Sarker

Department of Computer Science & Engineering, Chittagong University of Engineering & Technology, Chattogram, Bangladesh

CONTENTS

RECently, spam emails have become a significant problem with the expanding usage of the Internet. It is to some extend obvious to filter emails. A spam filter is a system that detects undesired and malicious emails and blocks them from getting into the users' inboxes. Spam filters check emails for something "suspicious" in terms of text, email address, header, attachments, and language. In this work, we have used different features such as word2vec, word n-grams, character n-grams, and a combination of variable length n-grams for comparative analysis. Different machine learning models such as support vector machine (SVM), decision tree (DT), logistic regression (LR), and multinomial naive Bayes (MNB) are applied to train the extracted features. We use different evaluation metrics such as precision, recall, f1-score, and accuracy to evaluate the experimental results. Among them, SVM provides 97.6 % of accuracy, 98.8% of precision, and 94.9% of f1-score using a combination of n-gram features.

14.1 INTRODUCTION

In recent years, web security has become one of the most critical issues. Most of our daily services have started using the internet, mobile computing, and electronic media. Email is one of the mediums for communication and increases in volume with the increasing use of the internet. Spamming is one of the most straightforward attacks in email messaging. Besides, users frequently receive annoying spam messages and malicious phishing messages by subscribing to different websites, products, services, catalogs, newsletters, and other types of electronic communications [1, 8]. In some cases, spam email is produced by mass-mailing viruses or Trojan horses. According to the China Anti-Spam Alliance's new survey on data, a typical Internet user receives 35 emails per week on average among which, 41% are spam emails. The presence of such spam messages wastes time as well as bandwidth on internet connections. Furthermore, they are often associated with offensive content and spread computer viruses. Due to these problems, cyber specialists are devoted to developing accurate spam detection in digital communication.

Moreover, there are many solutions for filtering spam, e.g., the blacklist and white-list filtering techniques, decision tree based approaches, email address based approaches, and machine learning based methods. The majority of them rely heavily on text analysis of the content of an email. As a result, there is a growing demand for effective anti-spam filters that automatically identify and remove spam messages or alert users to possible spam messages. However, spammers always investigate the loopholes of existing spam filtering techniques. They have introduced a new design for spreading spam emails in a wide range. Therefore, the existing system does not function against them. Tokenization attacks sometimes mislead spam filtering by adding extra spaces. Therefore, email contents need to be structured [14]. Moreover, in spite of having the highest accuracy in machine learning based spam email detection [3, 17], false positives (precision of 92.9%) are an issue due to one-shot detection of email threats. Addressing the false positive issues and changes in various attack designs, the stop words and other unwanted information are removed from the texts for further analysis in our proposed approach. After pre-processing, these texts go through numerous feature extraction methods, such as word2vec, word n-gram, character n-gram, and a combination of variable length n-grams. Different machine learning techniques such as the support vector machine (SVM), decision tree (DT), logistic regression (LR), and multinomial naive Bayes (MNB) [15] are applied on these matrices to classify the emails. The primary contributions of this chapter are the following:

(i) Create a content-based spam filter that can classify spam and ham emails.

(ii) Analyze numerous feature extraction methods, such as word2vec, word n-gram, character n-gram, and a combination of variable length n-grams.

(iii) Evaluate the performance of numerous experiments and achieve the best performance using support vector machine (SVM), decision tree (DT), logistic regression (LR) and multinomial naive Bayes (MNB) with proper features.

The rest of the chapter is organised as follows. Sect. 14.2 discusses the related works. The proposed approach for detecting email spam is presented in Sect. 14.3, the results and analysis are discussed in Sect. 14.4, and finally, the chapter is concluded in Sect. 14.5.

14.2 RELATED WORKS

Liu et al. [2] proposed a spam filtering method for emails based on their content. Their proposed technique is divided into two phases: training and classification. The extracted keywords from individual users' emails are compared against a spam and ham keywords corpus and achieved an overall accuracy of 92.8% and precision of 84.6%. Gaurav et al. [3] suggested a spam mail detection system for detecting spam emails based on the document labeling concept and applied three algorithms for email classification: Multinomial naive Bayes, Decision Tree, and Random Forest. The Random Forest technique achieved the highest accuracy of 92.97% and precision of 92.9% among these classifiers. Kanaris et al. [4] introduced a low-level data-based spam detection approach. Instead of using the bag-of-words approach to extract features, they employed character n-grams to create a bag-of-character-n-grams. Kiliroor et al. [5] suggested a model for detecting unwanted or unsolicited messages from the users' walls on online social networking sites, which had an accuracy of 91.18%. Weimiao Feng et al. [6] proposed a method based on SVM-NB. The SVM method is used to split training samples into different groups and to find dependent training samples. Moon et al. [7] proposed a spam mail filtering system based on n-gram indexing to support vector machines. They practiced with emails obtained from various users and performed the filtering procedure with the SVM classifier. Kaur et al. [8] proposed a spam detection technique using N-gram analysis and machine learning techniques. The N-grams that are built are used to predict unlabeled data.

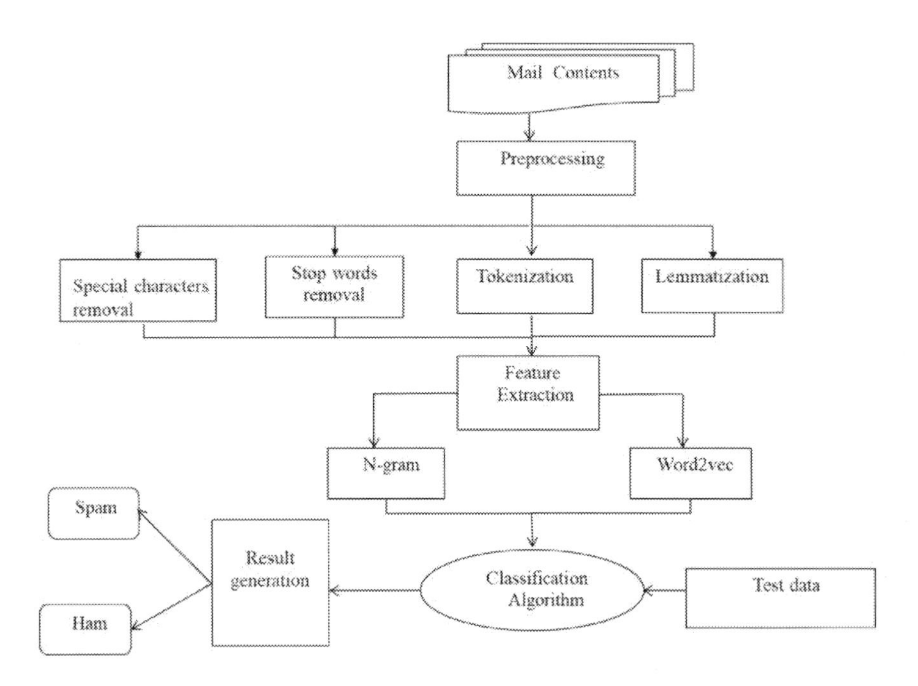

Figure 14.1 Steps of proposed methodology for content-based spam email detection.

Ahmad et al. [9] proposed a method that achieved 96% accuracy, in which an optimal subset of features is chosen for the learning process and a support vector classifier is used to classify. Sarker et al. [16] performed an effectiveness analysis of machine learning security modeling with optimal features on a broad scale. Nayak et al. [10] proposed a method for spam email detection that employs a hybrid bagging approach as a feature and combines the Naive Bayes and Decision Tree machine learning algorithms as classifiers, which achieve overall 88.12% of accuracy. Sheu et al. [11] proposed a method concentrating on email header analysis using a decision tree classifier to search for spam association guidelines at first. Next, an effective systematic filtering process is generated based on these association laws. Chen et al. [12] proposed a systematized spam filtering method based on a decision tree data mining methodology to evaluate spam association rules and apply these rules to create an effective spam filtering method. This method provides precision of 96%. Kumar et al. [13] proposed an approach that verifies the email header and URL as well as analyzes the body texts using different rules. They also employed a Bayesian classifier and the Apriori algorithm to classify files and attachments. S.A., et al. [17] proposed a framework working with email header features for email spam detection by analyzing two email datasets. The Support Vector Machine was used to classify email, which provides 88.80% accuracy.

14.3 METHODOLOGY

The proposed system for spam email detection is depicted in Figure 14.1. It has four phases: preprocessing, feature extraction, training, and prediction. Several preprocessing steps are performed before extracting features from the text. Feature extraction techniques are used to extract features from the preprocessed texts. In the training phase, these extracted features are used to train machine learning classifiers, such as support vector machine (SVM), decision tree (DT), logistic regression (LR), and multinomial naive Bayes (MNB). Finally, in the prediction stage, the contents of the emails are predicted as spam or ham.

14.3.1 Preprocessing

The *preprocessing* step involves eliminating inconsistencies and mistakes from raw data to make it more understandable. As a result, we must preprocess our data before feeding it into our model. Consider the following email:

"hello! We want to make localized version of the software...."

This email can be pre-processed in the following manner:

14.3.1.1 Special character removal

Each text is stripped of special characters such as (,=,>), numbers, and punctuation. After removal of those special characters from the content of the above email, the text would become *"hello we want to make localized version of the software"*.

14.3.1.2 Stop words removal

Words like "the," "an," "this," "a," etc., are not needed for recognizing spam or ham emails and hence those words are excluded. After removing the stop words, the text of the given email becomes *"want make localized version software"*.

14.3.1.3 Tokenization

Tokenization is the process of breaking down a large amount of text into smaller tokens. Tokenization provides a list of words such as *"want", "make", "localized", "version" and "software"* for the above given email.

14.3.1.4 Lemmatization

Lemmatization generally aims to eliminate only inflectional endings and restore the lemma, which is the base or dictionary form of a phrase. After lemmatizing the text of the email, we get words like *"want", "make", "localize", "version" and "software"*.

14.3.2 Feature extraction

The feature extraction phase converts raw data into useful knowledge by reformatting, merging, and converting primary features into new ones. We have used the following feature extraction techniques:

Table 14.1 Word n-gram representation of an email.

Sentence	"We want to make localized version of the software"
Uni-gram	'we', 'want', 'to', 'make', 'localized', 'version', 'of',...
Bi-gram	'we want', 'want to', 'to make', 'make localized', 'localized version',...
Tri-gram	'we want to', 'want to make', 'to make localized', 'make localized version',...

Table 14.2 Character n-gram representation of an email.

Sentence	"We want to make localized version of the software"
Uni-gram	'w', 'e', 'w', 'a', 'n', 't', 't','o', 'm', 'a', 'k', 'e', 'l', 'o', 'c', 'a', 'l', 'i', 'z','e','d'
Bi-gram	'we', 'ew', 'wa', 'an', 'nt', 'tt', 'to', 'om', 'ma', 'ak', 'ke', 'el', 'lo', 'oc'
Tri-gram	'wew', 'ewa', 'wan', 'ant', 'ntt', 'tto', 'oma', 'mak', 'ake', 'kel', 'elo', 'loc', 'oca', 'cal'

14.3.2.1 N-gram

An n-gram is an n-tuple or set of n words or characters that follow one another. The number of consecutive terms that can be treated as one gram is indicated by the letter 'n'. Machine learning algorithms cannot access raw text. We have then applied the word n-grams, character n-grams, and a combination of variable length n-grams to gain a better understanding of the sentences. The texts are not in structured forms. So, we convert text into numerical vectors using TF-IDF[1].

Word n-gram: Word n-grams deal with tuples or group of words. The word n-gram representation of an email is shown in Table 14.1.

Character n-gram: A text that is represented by a sequence of characters is known as a character n-gram. Unlike word n-grams, character n-grams can detect a word's identification and possible neighbors and the word's morphological makeup. Table 14.2 shows the character n-gram representation of an email.

Combination of variable length n-grams: The variable length is not pre-defined by a combination of variable length n-grams. It can merge uni-grams and bi-grams, or tri-grams and five-grams. Table 14.3 shows the combination of variable length n-gram representation of an email.

[1] https://towardsdatascience.com/text-vectorization-term-frequency-inverse-document-frequency-tfidf-5a3f9604da6d

Table 14.3 Combination of variable length n-gram representation of an email.

Sentence	"We want to make localized version of the software"
uni-gram+bi-gram	'We','want','to','make','localized', 'we want', 'want to', 'to make',...
bi-gram+tri-gram	'We want', 'want to','to make','make localized', 'We want to', 'want to make',...

14.3.2.2 Word2vec

Word2vec uses a neural network model to learn word associations from a large corpus of text. This type of model can identify synonyms and suggest new terms for a sentence. Word2vec correlates each different word with a specific set of integers known as a vector, as the name indicates. We generate a bag of words model out of the entire corpus, with each word being a vector.

14.3.3 Training

The features gathered in the previous phase are used to train a machine learning model. Support vector machine (SVM), decision tree (DT), logistic regression (LR), and multinomial naive Bayes (MNB) are used to train the extracted features in our proposed approach. In the following subsections, we discuss these algorithms.

14.3.3.1 Support Vector Machine (SVM)

Support vector machine is a supervised machine learning model. It generalizes between two classes. The first goal of the SVM is to seek out a hyperplane that will distinguish between the 2 classes. The hyper-plane equation is shown in Eq. 14.1:

$$w.x_i + b = 0 \tag{14.1}$$

where w is the weight factor, b is the bias, and x is the feature vector of sample i.

14.3.3.2 Logistic Regression (LR)

Logistic regression works effectively with binary classification problems. The activation *sigmoid* function's mathematical equation that results in binary classification is shown below in Eq. 14.2:

$$F(z) = \frac{1}{1 - e^-z} \tag{14.2}$$

Now, in the above equation,

$$z = w_0 + w_1.x_1 + w_2.x_2 + ... + w_n.x_n \tag{14.3}$$

The model's co-efficient produced using Maximum Likelihood Estimation is w_0, w_1, w_2,..., w_n, and the features or independent variables are x_0, x_1, x_2,..., x_n in the preceding equation. Finally, the binary outcome likelihood is calculated using z in the previous equation, where the possibilities are separated into two categories based on the given information (x).

14.3.3.3 Decision Tree (DT)

The decision tree has two types of nodes: external and internal nodes. External nodes represent the decision class, while internal nodes have the features required for categorization. A top-down strategy was used to examine the decision tree, which split homogeneous data into subsets. Its entropy is calculated using Eq. 14.4, which defines sample homogeneity.

$$E(s) = \sum_{l=1}^{n} p_i log_2 p_i \tag{14.4}$$

The entropy of a sample in the training class is $E(S)$, and the probability of a sample in the training class is p_i. Entropy was used to determine the splitting consistency. During the split, all of the features are considered to identify the appropriate split for each node. Random state 0 controls the recombination of the features.

14.3.3.4 Multinomial Naive Bayes (MNB)

In Natural Language Processing (NLP), the multinomial naive Bayes algorithm is a common probabilistic learning method. It assesses the probability of each tag for each sample and returns the tag with the highest probability. The Bayes theorem, as established by Bayes, calculates the likelihood of an event occurring based on prior knowledge of the conditions involved. In Eq 14.5 the formula is shown.

$$P(A|B) = P(A) \times P(B|A)/P(A) \tag{14.5}$$

When a predictor B is already available, we evaluate the likelihood of sophistication A. $P(B)$ denotes the probability distribution of B, $P(A)$ denotes the prior probability of sophistication A, and $P(B|A)$ denotes the probability of predictor B given the probability of class A.

The trained classifier models are then used for predicting the text contents as spam or ham.

14.4 RESULT AND OBSERVATIONS

We have implemented all the experiments in an Intel Core i5 processor with GPU 8 GB RAM in Python 3.7 on a Jupyter notebook.

The spam or ham email dataset is collected from Kaggle [2], an online data publishing source. The dataset contains 5731 emails, of which 1369 emails are spam and 4362 emails are ham. The training set contains 80% of the total data, while the testing set only contains 20% of the total data.

[2]https://www.kaggle.com/balakishan77/spam-or-ham-email-classification

Table 14.4 Performance comparison among different feature extraction methods.

Word-n gram	Classifier	Precision(%)	Recall(%)	f1 score(%)	Accuracy(%)
bi-gram	SVM	98.2	82.6	89.7	95.4
	LR	97.1	72.8	83.2	92.9
	DT	79.0	94.2	85.9	92.5
	MNB	100	74.2	85.2	93.8
tri-gram	SVM	96.7	74.6	84.2	93.2
	LR	98.2	83.6	90.4	95.7
	DT	70.4	91.6	79.6	88.7
	MNB	100	74.2	85.2	93.8
Character-n gram	Classifier	Precision(%)	Recall(%)	f1 score(%)	Accuracy(%)
bi-gram	SVM	88.1	51.0	64.6	86.5
	LR	86.1	47.4	61.2	85.5
	DT	71.4	57.2	63.5	84.2
	MNB	100	74.2	85.2	93.8
tri-gram	SVM	97.5	85.5	91.1	95.9
	LR	96.8	78.6	86.7	94.2
	DT	81.5	80.0	80.8	90.8
	MNB	100	74.2	85.2	93.8
Combination of n-grams	Classifier	Precision(%)	Recall(%)	f1 score(%)	Accuracy(%)
uni-gram+ bi-gram	SVM	98.8	91.3	94.9	97.6
	LR	98.7	82.9	90.1	95.6
	DT	87.0	88.0	87.0	93.9
	MNB	100	78.2	87.8	94.7
bi-gram+five-gram	SVM	99.2	90.9	94.8	97.6
	LR	98.7	85.1	91.4	96.1
	DT	87.9	87.3	87.6	94.0
	MNB	56.5	99.3	72.0	89.4
Word2vec	Classifier	Precision(%)	Recall(%)	f1 score(%)	Accuracy(%)
	SVM	91.3	7.6	14.0	77.6
	LR	83.6	37.0	53.1	83.1
	DT	55.0	55.6	55.3	78.4
	MNB	68.9	69.4	51.7	68.9

We use various machine learning algorithms to assess the system's performance, including SVM, MNB, DT, and LR. Again, various feature extraction approaches are used to test a variety of models. Table 14.4 shows the performance evaluation of different feature extraction methods. In the word n-gram feature, the SVM classifier provides the highest accuracy of 95.4% and precision of 98.2% considering bi-grams. In contrast, the logistic regression classifier achieves the best performance of 95.7% accuracy and 98.2% precision considering tri-grams. Again, for character n-gram, the naive Bayes classifier provides the highest 93.8% accuracy and 100% precision considering bi-grams and SVM achieves the highest 95.9% accuracy and 97.5% precision considering tri-grams. We also combine variable-length n-gram features and find that the combination of (uni-gram+bi-gram) provides the highest accuracy of 97.6% and

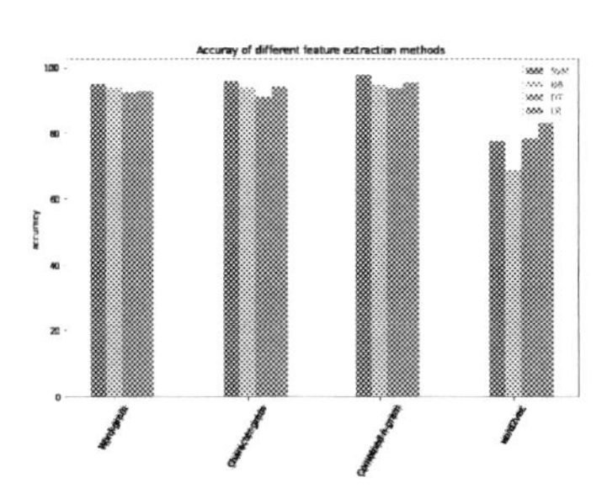

Figure 14.2 Accuracy comparison among different ML classifiers.

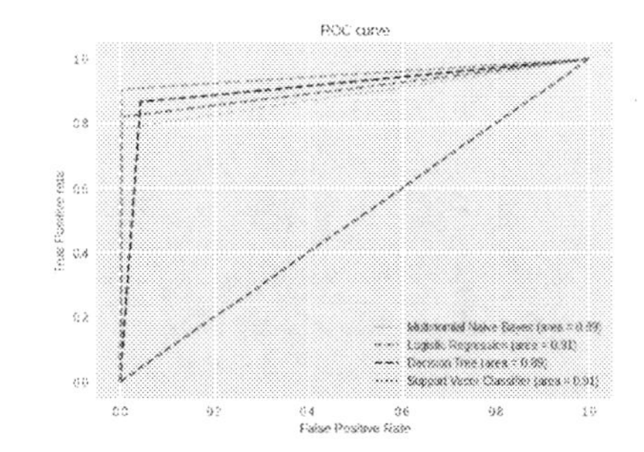

Figure 14.3 ROC curve for combination of variable length n-gram using various classification models.

Table 14.5 Comparison of our work with other benchmark works

Reference	Feature Extraction	Classifier	Accuracy	Precision
[3]	Document labeling	Random Forest	92.97%	92.90%
[17]	Header features	SVM	88.80%	-
Our method	Combination of Variable length n-gram	SVM	97.6%	98.8%

precision of 98.8% using the SVM classifier. For the Word2vec method, the logistic regression classifier provides the highest 83.1% of accuracy and 83.6% of precision.

We consider the uni-gram, bi-gram, and five-gram to investigate the features from email content. Using uni-grams, we alleviate the unwanted tokenization of white spaces. It decreases the possibility of spam considered to be ham (FN) or vice versa (FP) and provides the highest accuracy as shown in Figure 14.2. In conclusion, SVM has proven to be the best classifier and is effective in recognizing capabilities for our dataset due to robustness in high-dimensional features. With the combination of uni-grams and bi-grams, SVM achieves the highest accuracy of 97.6%, precision of 98.8%, and f1-score of 94.9%.

A receiver operating characteristic (ROC) curve shows how well a classification model performs over various classification thresholds. We have drawn the ROC curve for detecting spam emails for the combination of uni-gram and bi-gram using different classifiers as shown in Figure 14.3. Logistic regression and the SVM classifier perform better from the perspective of the area under the ROC curve, and it is 0.91 in both cases.

Further, we also compare our proposed method with some benchmark works of [3, 17] as shown in Table 14.5. Our proposed method outperforms the two benchmark methods with an accuracy of 97.6% and precision of 98.8% using the SVM classifier.

14.5 CONCLUSION

Accurate spam detection is an integral part of email communication. Despite accurate detection of spam [3, 17], the false positive rate is also an issue. To address this, we present a content-based spam email detection approach. We use multinomial naive Bayes, logistic regression, support vector machine, and decision tree classifiers for learning the various features from the contents of emails. For comparative research, we use word n-grams (bi-gram, tri-gram), character n-grams (bi-gram, tri-gram), the combination of variable length n-grams (uni-gram and bi-gram, bi-gram and five-gram), and word2vec features. Among them, SVM achieves the best performance of 97.6% accuracy, 98.8% precision, and 94.9% f1-score for the combination of variable length n-grams (uni-gram and bi-gram). In the future, we can extend our work by analyzing the features using context-based machine learning.

Bibliography

[1] Sarker, Iqbal H., Md Hasan Furhad, and Raza Nowrozy. (2021). AI-driven cybersecurity: An overview, security intelligence modeling and research directions, *SN Computer Science*, 2.3, 1–18

[2] Liu, P., & Moh, T. (2016). Content Based Spam E-mail Filtering. *2016 International Conference on Collaboration Technologies and Systems (CTS)*, 218–224.

[3] Gaurav, Devottam and Tiwari, Sanju Mishra and Goyal, Ayush and Gandhi, Niketa and Abraham Ajith.(2020). Machine Intelligence-based Algorithms for Spam Filtering on Document Labeling, *Soft Computing - A Fusion of Foundations, Methodologies and Applications, Springer*, 9625–9638

[4] Kanaris, I., Kanaris, K., Stamatatos, E. (2006). Spam Detection Using Character N-Grams. *In: Antoniou, G., Potamias, G., Spyropoulos, C., Plexousakis, D. (eds) Advances in Artificial Intelligence. SETN 2006. Lecture Notes in Computer Science*, vol 3955. Springer, Berlin, Heidelberg.

[5] Kiliroor, C.C., Valliyammai, C. (2019). Social Context Based Naive Bayes Filtering of Spam Messages from Online Social Networks. *In: Nayak, J., Abraham, A., Krishna, B., Chandra Sekhar, G., Das, A. (eds) Soft Computing in Data Analytics . Advances in Intelligent Systems and Computing*, vol 758. Springer, Singapore.

[6] Feng, W., Sun, J., Zhang, L., Cao, C.,& Yang, Q. (2016). A support vector machine based naive Bayes algorithm for spam filtering. *2016 IEEE 35th International Performance Computing and Communications Conference (IPCCC)*, 1–8.

[7] Moon, J., Shon, T., Seo, J., Kim, J., Seo, J. (2004). An Approach for Spam E-mail Detection with Support Vector Machine and n-Gram Indexing. *Computer and Information Sciences - ISCIS 2004. ISCIS 2004. Lecture Notes in Computer Science*,vol. 3280, Springer, Berlin, Heidelberg.

[8] Kaur, S. (2019). Spam Detection using N-gram Analysis and Machine Learning Techniques.[online] Available at: https://www.semanticscholar.org/paper/

Spam-Detection-using-N-gram-Analysis-and-Machine-Kaur [Accessed 28 Feb. 2020]

[9] Ahmad, S.B.S., Rafie, M. & Ghorabie, S.M.(2021). Spam Detection on Twitter Using a Support Vector Machine and Users' Features by Identifying Their Interactions. *Multimed Tools Appl.*, 80, 11583–11605

[10] Nayak, Rakesh and Jiwani, Salim Amirali and Rajitha, B. (2021). Spam email detection using machine learning algorithm, *Materials Today: Proceedings, Elsevier*

[11] Sheu, J., Chu, K., Li, N., & Lee, C. (2017). An efficient incremental learning mechanism for tracking concept drift in spam filtering. *Public Library of Science San Francisco, CA USA*, 12.

[12] Sheu, J., Chen, Y., Chu, K., Tang, J., & Yang, W. (2016). An intelligent three-phase spam filtering method based on decision tree data mining. *Secur. Commun. Networks*, 9, 4013–4026.

[13] Kumar, S., Gao, X., Welch, I., & Mansoori, M. (2016). A Machine Learning Based Web Spam Filtering Approach. *2016 IEEE 30th International Conference on Advanced Information Networking and Applications (AINA)*, 973–980.

[14] Sarker, Iqbal H.(2021). Data science and analytics: An overview from data-driven smart computing, decision-making and applications perspective, *SN Computer Science*, 1–22

[15] Sarker, Iqbal H.(2021). Machine learning: Algorithms, real-world applications and research directions, *SN Computer Science*, 2.3, 1–21

[16] Sarker, Iqbal H.(2021). CyberLearning: Effectiveness analysis of machine learning security modeling to detect cyber-anomalies and multi-attacks, *Internet of Things*, 14, 100393

[17] Khamis, S.A., Foozy, C.F.M., Aziz, M.F.A., Rahim, N. (2020). Header Based Email Spam Detection Framework Using Support Vector Machine (SVM) Technique. *In: Ghazali, R., Nawi, N., Deris, M., Abawajy, J. (eds) Recent Advances on Soft Computing and Data Mining. SCDM 2020. Advances in Intelligent Systems and Computing*, vol 978. Springer, Cham.

CHAPTER 15

AI Poet: A Deep Learning-based Approach to Generate Artificial Poetry in Bangla

Hasan Murad

Dept. of CSE, University of Asia Pacific, Dhaka, Bangladesh

Rashik Rahman

Dept. of CSE, University of Asia Pacific, Dhaka, Bangladesh

CONTENTS

Natural Language Processing (NLP) is one of the fast-growing subfields of Artificial Intelligence (AI) and text generation is an advanced application of NLP. Though a significant amount of effort has been devoted to developing text generating models in different languages, far too little has yet been done to generate artificial text in Bangla due to the lack of a sufficient available text dataset. In our research work, we have created a noble dataset with a collection of poems written by our national poet Kazi Nazrul Islam. We have designed a Bidirectional LSTM based model architecture for Bangla text generation. After necessary pre-processing on our noble dataset, we have trained our model and achieved significant promising results on artificial text generation. Our trained model can generate meaningful artificial poetry with a writing pattern similar to poet Kazi Nazrul Islam.

15.1 INTRODUCTION

Bangla is the official and national language of the people's republic of Bangladesh. Bangla literature is a vast sector that is still untouched by Natural Language Processing (NLP) which is one of the most important subfields of Artificial Intelligence (AI).

Text generation is one of the most recent and exciting research topics in NLP. Significant efforts have been devoted to applying NLP in text generation [1, 2, 3, 4]. However, far too little has yet been done to generate artificial text in Bangla due to the lack of a sufficient available text dataset.

In a recent paper, [1] designed a deep learning based model for Bangla text generation and trained the model using a public dataset created from Wikipedia. In another paper, [10] designed a recurrent model to generate artificial text in Bangla. They have also trained their model using a public dataset on Bangla news.

However, no recent work has been found for generating artificial Bangla literature using a deep learning based model. Moreover, training deep learning based models using a random dataset from Wikipedia does not provide satisfactory results for text generation.

Therefore, the objective of our research work is to develop a domain-specific noble dataset with a collection of poems written by our national poet Kazi Nazrul Islam and train a deep learning based model that can generate meaningful artificial poetry with a similar writing pattern as poet Kazi Nazrul Islam.

As there is no existing dataset on Bangla poems, we have developed a large noble dataset with a collection of poems of Kazi Nazrul Islam. In the pre-processing step, we have generated sequences of words from the poems using n-grams. Finally, we have applied word embedding to vectorize our dataset. For Bangla text generation we have designed a Bidirectional LSTM based model architecture. In Section 15.3, we explain our model architecture with necessary details. Finally, we train our model on the prepared dataset.

Our model can predict the next possible word from a given initial word sequence as input. Then, we append the predicted word with the initial sequence to create a new sequence. Our model again predicts the next possible word from the

Table 15.1 Generated artificial poem

Given Sequence	Generated Poem
ঠাই নাই ঠাই নাই ছোট সে তরী নাই কেউ	ঠাই নাই ঠাই নাই ছোট সে তরী নাই কেউ আমি গোপন প্রিয়ার চকিত চাহনি ছল করে দেখা অনুখন যেন খোঁপায় গুঁজে নেও ভয় কি সখি এ বর্ষা ঝরা এমনি প্রাতে জাগবে হঠাৎ চমকে কাহার বৃন বসি হয়ে তৃষিত জলধি লাখে কিসের আশায় করছে ধারা গ্রহ শশী ধরার ফুটিল বিপ্লব অহিদল খুঁজে ফেরে ডেরা সন্ধ্যায
এমন দিনে কোথায় তোরে আড়াল দিয়ে রাখি?	এমন দিনে কোথায় তোরে আড়াল দিয়ে রাখি? মরণ প্রাণ হে ব্যথা সাধক অশ্রু সরস্বতী কর্ণে তুমি কুরুবক হে পথিক বন্ধু হে সিন্ধু হে বিরহী বন্ধু মোরা – করে টলমল এক জ্বালা এক ব্যথা নিয়া ব্যথা নিয়া এক ভাগ বাকী সুরা নাই পাত্র হাতে দেখব এবার জগৎটাকে
গান ফুরালো যাব যবে গানের কথাই মনে	গান ফুরালো যাব যবে গানের কথাই মনে রবে পাখী তখন থাকবো না থাকবে পাখীর স্বর উড়ব আমি কাঁদবে তুমি ব্যথার বালুচর তোমার পারে বাজল কখন আমার পারের ঢেউ অজানিতা কেউ জানে না জানবে না ।
তোমার পারে বাজল কখন আমার পারের	তোমার পারে বাজল কখন আমার পারের ঢেউ অজানিতা কেউ জানে না জানবে না ক কেউ সে নর ও নারীতে সমান মিশিয়া রহে আবার যদি লোভী আসবে জোয়ার দুলবে তরী রঙ্গে সেই তরীতে হয়ত কেহ থাকবে তুমি নিয়া নেশা গানের সাত সুর আলো আঞ্ছে মানিক

new sequence. This process is repeated until we get a full poem with the predicted word sequence. In Section 15.5, we present some of the generated artificial poems in Table 15.1.

The main contributions of our research works are:

- We have developed a noble dataset with a collection of poems written by our national poet Kazi Nazrul Islam.
- We have designed a Bidirectional LSTM based model architecture to generate meaningful artificial poetry with a writing pattern similar to poet Kazi Nazrul Islam.
- We have deployed different types of recurrent layers such as RNN, Bidirectional RNN, LSTM, and Bidirectional LSTM in our model architecture and compared the results in Bangla text generation.

We organize the rest of the chapter in the following order. Section 15.2, describes related background knowledge and some of the relevant previous research works. In

Section 15.3, we present our proposed approach. Section 15.4 covers necessary implementation details. We present the result and evaluation of our research in Section 15.5. Finally, Section 15.6 concludes our research work and discusses possible future directions.

15.2 BACKGROUND AND LITERATURE REVIEW

Related background knowledge of deep learning based approaches in NLP and some of the relevant previous research work in the field of text generation are presented in this section.

15.2.1 Related Terminologies

In computer vision tasks such as object detection[11, 12], and segmentation[13], YOLOv4 and Convolutional Neural Network models are used extensively. In addition, the DBSCAN algorithm is widely used by researchers to solve clustering issues [14]. However, these algorithms are inappropriate for NLP tasks. NLP has been one of the most important tools in AI for language processing [5]. NLP has solved challenging problems like text classification, machine translation, text summarization, etc. [6, 7, 8, 15]. Deep neural network architectures like RNN(Recurrent Neural Network), LSTM(Long Short-Term Memory), and GRU(Gated Recurrent Unit) have enabled deep learning models to identify the sequential information which is necessary to solve NLP tasks. Text generation is one of the most exciting research topics in NLP.

15.2.2 Existing Works

Artificial text generation using deep learning is one of the most existing research works among practitioners [1, 2, 3, 4]. We discuss a set of previous works on text generation using NLP.

In a recent paper, Iqbal et al. [1] discussed several deep learning based model architectures for the text generation using RNN, LSTM, Bi-LSTM, and GAN for the English language. In another paper, Subramanian et al. [2] designed an adversarial model to generate a high-level sentence outline. Using the generated outline, they proposed a model to generate word sequences. Hou et al. [3] developed a deep learning based model for story generation using the sequence-to-sequence model.

However, far too little has been done to generate artificial text in Bangla due to the lack of sufficient available Bangla text datasets. In a recent paper, Abujar et al. [9] proposed a bi-directional RNN based model to generate text where they trained the model using a public dataset created from Wikipedia. In another paper, Islam et al. developed a sequence-to-sequence model using LSTM for text generation [10] where they also trained their model using a public dataset on Bangla news.

15.2.3 Limitations of the Existing Works

Most of the previous works for Bangla text generation used random Bangla text datasets collected from Wikipedia, which is not so effective in text generation. A

domain specific text dataset like that of a particular poet has not been found in Bangla for text generation.
Therefore, we have developed our noble text dataset with a collection of poems written by Bangladesh's national poet Kazi Nazrul Islam and designed a bi-directional LSTM network to generate meaningful artificial poetry with a writing pattern similar to poet Kazi Nazrul Islam.

15.3 PROPOSED APPROACH

In this section, we discuss our dataset, the data pre-processing pipeline, and the proposed deep learning based model architecture for artificial Bangla text generation.

15.3.1 Dataset Creation

We have created a collection of 100 poems written by Bangladesh's national poet Kazi Nazrul Islam from both online sources such as different websites providing a digital version of Bangla poems and offline sources such as Bangla poetry books. We have conducted the necessary data cleaning as some of the poems are collected from different websites. We have generated the corpus by extracting sentences from poems. We also divide our corpus into training (80%) and testing (20%) corpus.

15.3.2 Data Pre-processing

A deep learning based model cannot process raw text data. We have to convert textual data into numeric data by replacing each word with a unique number. Firstly, we tokenize each sentence in the training corpus, create a vocabulary set with all unique words, and assign a unique number for each word. However, to handle unknown words on the testing corpus, we use a special token called the Out of Vocabulary or OOV token where the assigned number of the OOV token is 1. For both the training and testing corpus, we replace each word of each sentence with the corresponding unique number and generate a sequence of numbers for each sentence. To increase the volume of our training corpus, we generate all the n-grams for each sequence. We add padding with zeros (0) for each sequence to make each sequence the max sequence length. Finally, we apply one hot encoding before providing each sequence as input in our model.

15.3.3 Model Architecture Design

Model architecture is presented in Figure 15.1. The first layer of our model architecture is the Input layer, which takes each sequence as input. The next layer is the Embedding layer which incorporates word embedding by replacing each word vector. The next layer is a recurrent layer that takes the output of the Embedding layer and determines the sequential information. We can use different types of recurrent layers such as RNN, bi-directional RNN, LSTM, and bi-directional LSTM. In Section 15.5, the bi-directional LSTM model provides a high BLEU score. The main strength of using bi-directional LSTM is that it can process the text in both forward and

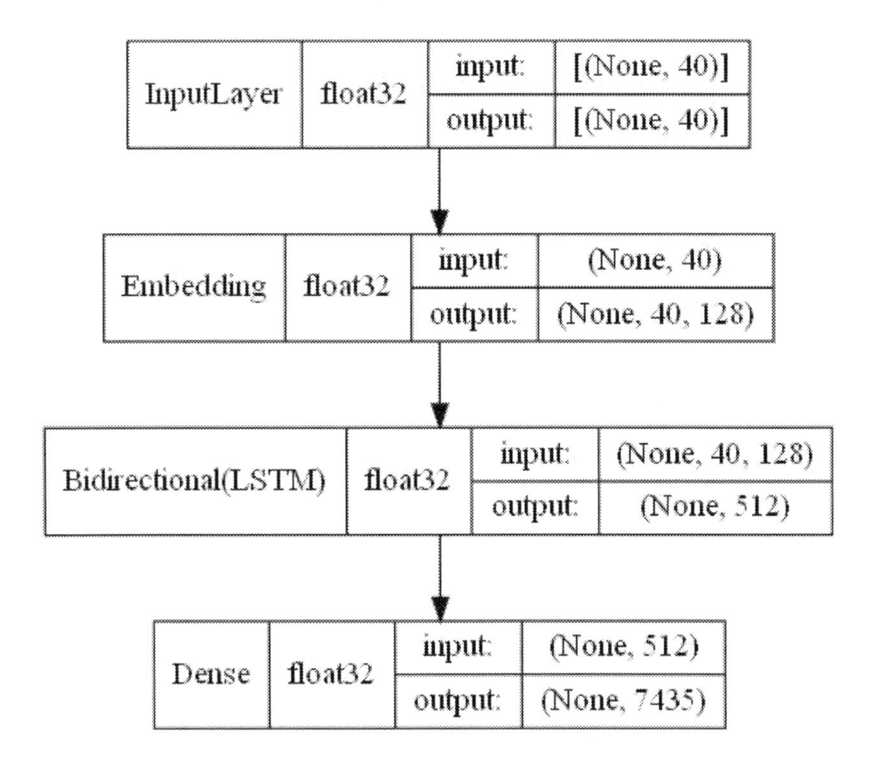

Figure 15.1 Model Architecture

backward directions and generates high-quality feature vectors. Using the output feature vector, the final Dense layer with softmax activation generates probabilities for the next possible predicted words.

15.4 IMPLEMENTATION

A brief overview of development tools, parameter setting, and the implementation details of our deep learning based model architecture for artificial Bangla text generation is presented in this section.

15.4.1 Development Tools

We implement our model architecture in Tensorflow 2.0. Tensorflow is a free and open-source machine learning framework that provides all the necessary libraries to design, develop, and deploy machine learning models. Keras is a high-level TensorFlow API for deep learning. Keras has all the essential neural layers like the Dense layer, RNN layer, LSTM layer, or bi-directional LSTM layer already built in. We design our proposed model architecture using the built layers available in the Keras API.

15.4.2 Pre-processing Pipeline

The pre-processing pipeline is implemented using a built-in library in Keras called Tokenizer. A tokenizer is a text pre-processing library that provides necessary functionalities for building a vocabulary set with all the unique words in the training corpus, replacing each word of each sentence with a unique number to create a sequence with numbers for text data, and finally padding each sequence to the max sequence length. Moreover, we can easily handle the out of vocabulary (OOV) words using the Tokenizer library.

15.4.3 Model Architecture Implementation

To implement the model architecture, we use the Sequential model, where all layers of the model are stacked. The first layer of the model architecture is the Input layer, which takes input with length 40. The next layer of our model is the Embedding layer. Keras provides a built-in layer called the Embedding Layer for word embedding. The output dimension of the Embedding layer is set to 128. The next layer is the bi-directional LSTM layer. In Keras, we implement the bi-directional LSTM layer using two separate layers named the bi-directional layer and LSTM layer. The return sequence parameter is False for the LSTM layer in the default setting. Therefore, the output of the bi-directional LSTM layer is a feature vector of size 512 at the last LSTM block. Finally, we add a dense layer with a softmax activation function as the last layer of our model to generate probabilities for the next possible predicted words.

15.5 RESULTS

In this section, we discuss a brief overview of the parameter setting used while training our model and present all the results of our research work.

15.5.1 Training Results

After training our model for 200 epochs, we achieve an accuracy of 90.05% and the loss is less than 0.05. In Figure 15.2, we observe the loss vs epoch curve where our model starts to converge right after 180 epochs. Moreover, the accuracy vs epoch curve also shows a similar result in Figure 15.3.

15.5.2 Parameter Setting

Categorical cross-entropy and Adam are used as the loss function and optimization algorithm, respectively. During the training phase, the batch size and learning rate were set to 64 and 0.01. We train the model for 200 epochs before convergence.

15.5.3 Environment Setting

We trained our model on the Google Colab platform. After selecting the GPU runtime environment, we used the following hardware configuration: Tesla T4 16GB GPU, Intel Xeon Processors @ 2.2Ghz CPU, 13GB of RAM and, 30GB of storage space.

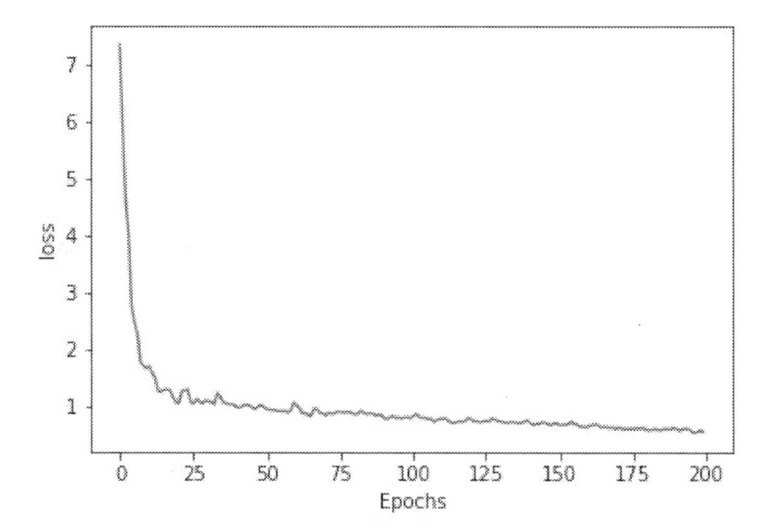

Figure 15.2 Loss vs Epoch

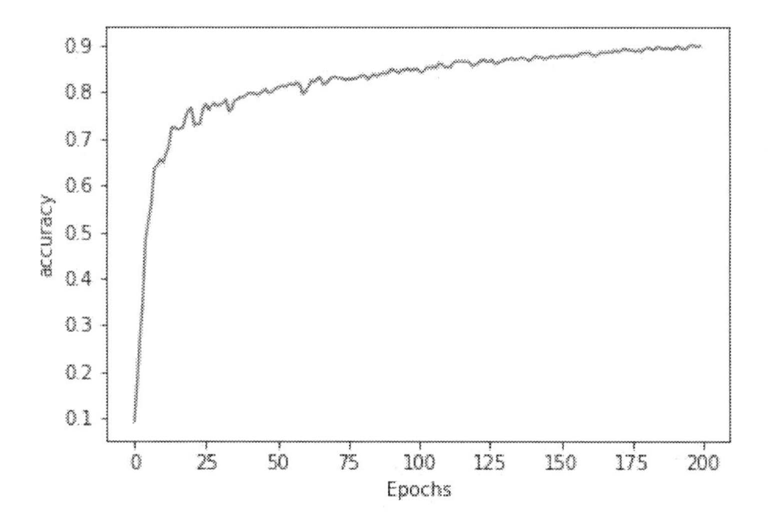

Figure 15.3 Accuracy vs Epoch

15.5.4 Evaluation

To evaluate our model, we provide a sample Bangla poem sequence as input in our trained model. Our model generates a meaningful artificial poem that has a writing rhythm similar to poet Kazi Nazrul Islam. In Table 15.1, we have provided some examples of generated Bangla poems using our trained model.

The BLEU score is an evaluation metric of the text generation model. The BLEU score compares the generated text with the reference text and provides a score. BLEU scores of the model for different numbers of units in bi-directional LSTM are shown in Table 15.2.

Table 15.2 BLEU score of the bi-directional LSTM model

No. of units	BLEU-1 score	BLEU-2 score
32	64.5	42.4
64	63.4	40.7
128	65.7	44.1

15.5.5 Limitations of Our Work

There is still scope for further improvement of the accuracy of our model by using more sophisticated complex model architecture. Moreover, the generated artificial poem has no punctuation as we have ignored punctuation while pre-processing our dataset.

15.6 CONCLUSION

In this research, we have developed a noble dataset with a collection of poems written by our national poet Kazi Nazrul Islam. We have designed a deep learning based model architecture using bi-directional LSTM for artificial Bangla text generation. After necessary data pre-processing and vectorizing the dataset with word embedding, we have trained our model and achieved high accuracy. If we provide a simple Bangla sentence as input in our trained model, it can generate meaningful artificial poetry with a writing pattern similar to poet Kazi Nazrul Islam. In the future, we can deploy a more complex model architecture for better performance in generating artificial Bangla literature.

Bibliography

[1] Iqbal, T., & Qureshi, S. (2020). The survey: Text generation models in deep learning. *Journal of King Saud University-Computer and Information Sciences.*

[2] Subramanian, S., Rajeswar, S., Sordoni, A., Trischler, A., Courville, A., & Pal, C. (2018, December). Towards text generation with adversarially learned neural outlines. In *Proceedings of the 32nd International Conference on Neural Information Processing Systems* (pp. 7562–7574).

[3] Hou, C., Zhou, C., Zhou, K., Sun, J., & Xuanyuan, S. (2019, October). A survey of deep learning applied to story generation. In *International Conference on Smart Computing and Communication* (pp. 1–10). Springer, Cham.

[4] Sellam, T., Das, D., & Parikh, A. P. (2020). BLEURT: Learning robust metrics for text generation. *arXiv preprint arXiv:2004.04696.*

[5] Khan, W., Daud, A., Nasir, J. A., & Amjad, T. (2016). A survey on the state-of-the-art machine learning models in the context of NLP. *Kuwait Journal of Science, 43*(4).

[6] Lai, S., Xu, L., Liu, K., & Zhao, J. (2015, February). Recurrent convolutional neural networks for text classification. In *Twenty-ninth AAAI Conference on Artificial Intelligence.*

[7] Singh, S. P., Kumar, A., Darbari, H., Singh, L., Rastogi, A., & Jain, S. (2017, July). Machine translation using deep learning: An overview. In *2017 International Conference on Computer, Communications and Electronics (comptelix)* (pp. 162–167). IEEE.

[8] Song, S., Huang, H., & Ruan, T. (2019). Abstractive text summarization using LSTM-CNN based deep learning. *Multimedia Tools and Applications, 78*(1), 857-875.

[9] Abujar, S., Masum, A. K. M., Chowdhury, S. M. H., Hasan, M., & Hossain, S. A. (2019, July). Bengali text generation using bi-directional RNN. In *2019 10th International Conference on Computing, Communication and Networking Technologies (ICCCNT)* (pp. 1–5). IEEE.

[10] Islam, M. S., Mousumi, S. S. S., Abujar, S., & Hossain, S. A. (2019). Sequence-to-sequence Bangla sentence generation with LSTM recurrent neural networks. *Procedia Computer Science*, 152, 51–58.

[11] Rahman, M., Rahman, R., Supty, K. A., Sabah, R. T., Islam, M. R., Islam, M. R., & Ahmed, N. (2022, March). A real time abysmal activity detection system towards the enhancement of road safety. In *2022 2nd International Conference on Innovative Research in Applied Science, Engineering and Technology (IRASET)* (pp. 1–5). IEEE.

[12] Rahman, R., Rakib, A. F., Rahman, M., Helaly, T., & Pias, T. S. (2021, November). A real-time end-to-end Bangladeshi license plate detection and recognition system for all situations including challenging environmental scenarios. In *2021 5th International Conference on Electrical Engineering and Information & Communication Technology (ICEEICT)* (pp. 1–6). IEEE.

[13] Rahman, R., Pias, T. S., & Helaly, T. (2020, October). Ggcs: A greedy graph-based character segmentation system for Bangladeshi license plate. In *2020 4th International Symposium on Multidisciplinary Studies and Innovative Technologies (ISMSIT)* (pp. 1–7). IEEE.

[14] Amiruzzaman, M., Rahman, R., Islam, M. R., & Nor, R. M. (2021, November). Evaluation of DBSCAN algorithm on different programming languages: An exploratory study. In *2021 5th International Conference on Electrical Engineering and Information & Communication Technology (ICEEICT)* (pp. 1–6). IEEE.

[15] Maraz, M. R. J., Rahman, R., Hasnain, M. M. U., & Murad, H. (2021, September). A cross-platform blood donation application with a real-time, intelligent, and rational recommendation system. In *2021 International Conference on Electronics, Communications and Information Technology (ICECIT)* (pp. 1–4). IEEE.

CHAPTER 16

Document Level Comparative Sentiment Analysis of Bangla News Using Deep Learning-based Approach LSTM and Machine Learning Approaches

Nuren Nafisa

Department of Computer Science and Engineering, Chittagong University of Technology and Engineering, Chattogram, Bangladesh

Sabrina Jahan Maisha

Department of Computer Science and Engineering, BGC Trust University Bangladesh, Chandanaish Chattogram, Bangladesh

Abdul Kadar Muhammad Masum

Department of Computer Science and Engineering, International Islamic University Chittagong, Chattogram, Bangladesh

CONTENTS

LIVING in the age of digitalization, people are empowered to easily express themselves, but they also be easily manipulated by information. Online news is one of the powerful mediums that can influence people's thoughts and ideas. Sensitive news may create inconvenience and also in some cases for certain age groups or people, it can suffer people's mental health. Fewer research works have been done in extracting sentiments from newspapers in a morphologically rich language like Bangla. In this chapter, our aim is to automatically analyze bipolar sentiments (i.e., positive or negative) from news articles and comparatively evaluate the behavior of traditional supervised Machine Learning (ML) classifiers with respect to the deep learning approach Long Short-Term Memory (LSTM) on our Bangla News dataset. In this respect, the dataset is collected from different online news sources, and with the help of four annotators of our acquaintance, each news document is labelled. To build a model with ML, six supervised classifiers (i.e., Multinomial Naive Bayes (MNB), K-Nearest Neighbor (KNN), Random Forest (RF), Decision Tree (DT), Logistic Regression (LR) and Linear Support Vector Machine (SVM)) are analyzed along with the Count Vectorizer and TF-IDF transformer techniques. Evaluated with the percentage split method, it is seen that RF shows the best results with 80% accuracy. Conversely, another model is created employing the Word2Vec technique with the LSTM approach. With 100 epochs, this method revealed an accuracy of 83%. Our model has been able to perform well using LSTM rather than the RF approach. Although in both cases accuracy can be considered quite good, an increase of the dataset can stimulate the reliability of model.

16.1 INTRODUCTION

With the exponential increase of internet users, huge amounts of content are generated regularly on online platforms. News is one of the most powerful media read by people irrespective of age, caste, race and religion. As various types of posted writings can impact people mentally in a positive or negative manner, content of news must be handled in a sophisticated way. This data should be analyzed to extract the inherent sentiments of the writer. In this respect, Sentiment Analysis (SA) emerged as an effective tool to portray the referred meaning exhibited in such online news or any other social media content. Although, many works have been done in these field of SA in languages such as English, Chinese and so on. Few research works have been done so far in the Bangla language.

The work done in the Bangla language has been mostly done in the domain of reviews, tweets, blogs or comments. As very few standard datasets are available publicly, this work aims to build a Bangla news corpus. For this purpose, the corpus is collected from different news sources and with the help of four different audiences. Bangla news content have been annotated as negative and positive. Analyzing different research works in this field, it has been seen that machine learning (ML) algorithms have shown propitious results gaining popularity in most cases [1]. Considering this, in our proposed system, six supervised ML algorithms, SVM, DT, MNB, KNN, RF and LR, are applied to analyze the dataset. However, in various kinds of NLP tasks, especially for text analysis, deep learning based advanced model Long Short-Term Memory (LSTM) approaches have shown great success. Generally, texts carry inherent meanings of the context, and refer to it sequentially. With the feature of grasping the previous information, LSTM can perform well in long text-based data. This kind of feature can be well handled by LSTM because of its inner architectural mechanism. For this reason, our model has chosen LSTM to evaluate the dataset. Finally, a comparative study can be done to understand the behavior of the classifiers on the new Bangla news dataset as well as estimate an efficient model for the proposed work. Thus, the key contributions of our work can be outlined as follows:

- To build a large Bangla news corpus dataset
- To extract bi-polar sentiments using six traditional supervised ML algorithms
- Automatically analyze the sentiments from the corpus with the help of the deep learning-based approach LSTM
- Analyze the most suitable model for this domain among supervised ML and LSTM approaches

The chapter is organized as follows. In Section 16.2, literature review in this arena are illustrated. In Section 16.3, the dataset development process is discussed in a detailed manner. In Section 16.4, methodology of the whole proposed model is depicted, and every component related to the system is discussed. In Section 16.5, the behavior of the classifiers and the results obtained through experimentation are shown. In Section 16.6, the concluding expression derived from the proposed model is stated.

16.2 LITERATURE REVIEW

In this section, some of the significant works done in the study of SA in the Bangla language have been discussed.

16.2.1 SA in Bangla Language

Table 16.1 and Table 16.2, reviews of the works done so far in the Bangla language in textual context such as news, tweets and so on. Going through the notable research

Table 16.1 Literature Review of researches of SA on Bangla Language.

Authors	Methodology	Purpose	Dataset	Result/Findings
Chowdhury and Chowdhury, 2014 [2]	Using SVM and Maximum Entropy (MaxEnt), analysis is performed.	To detect the positive and negative sentiments from Bangla microblog texts	1300 Bangla tweets	This approach found 98% accuracy applying SVM model with unigrams and emoticons as features.
Hasan et al., 2016 [3]	In this work, deep recurrent neural network-LSTM has been applied.	To categorize Bangla and Romanized Bangla textual dataset into positive, negative and ambiguous and analyze it using deep recurrent neural network-based approach.	Collected 10,000 Bangla and Romanized Bangla text samples from Facebook, Twitter and some online news and product review portals	Considering categorical cross-entropy and cases of with and without ambiguous classification, the model achieved 55% and 78% accuracy respectively
Tabassum and Khan, 2019 [4]	Integrating unigram, POS tagging, negation handling with supervised ML classifier RF this model has been developed.	To establish an efficient model of Bangla texts in extracting positive and negative sentiments	1050 Bangla texts accumulated from Facebook and Twitter comments	With an error of 13%, this system obtained 85% accuracy through unigram, POS and negation techniques.

(*Continued on next page*)

Table 16.1 (*Continued*)

Authors	Methodology	Purpose	Dataset	Result/Findings
Wahid et al., 2019 [5]	Word embedding technique along with RNN-LSTM based approach has been implemented.	To identify the sentiments expressed in cricket comments as positive, negative and neutral	"ABSA" cricket comments dataset having 2979 data	More than 90% accuracy has been achieved in this case after 15 epochs with batch size 30 using LSTM approach.
Hasan et al., 2020 [6]	SVM, RF and deep learning algorithms Convolutional Neural Network (CNN), Fast-Text, BERT, DistilBERT and XLM-RoBERTa have been implemented in this work.	To evaluate various publicly available datasets on SA with the help of traditional and DL algorithms	ABSA: Cricket(2837) and Restaurant(1808), BengFastText(8420), SAIL(998), YouTube Comments(2796) and Social Media Posts(6570)	Results depicted that deep learning transformer-based model BERT and XLM-RoBERTa performed well with 72.9% among others.
Hussain et al. 2020 [7]	Extracting features with Count Vectorizer and TF-IDF, two supervised learning algorithms MNB and SVM have been used.	To identify the Bangla fake news in social media	About 2500 articles have been collected from 4 Bangla online news sources.	The performances showed that SVM achieved a higher accuracy of more than 90%.

Table 16.2 Data statistics of Bangla news corpus.

Type of News	Positive News	Negative News
Number	1048	2029
Total	3077	

works in the field of SA in the Bangla language, works have been done in analyzing tweets, reviews, or textual content. Most of the datasets are not accessible publicly and corpus size is also small. However, different supervised ML approaches and deep learning methods have been adopted in different contexts such as SVM, DT and so on. Thus, in our work, document analysis is done to propose the best model for the Bangla news corpus by comparing six popular supervised ML classifiers and an advanced deep learning-based approach LSTM, which has also shown significant results in textual analysis.

16.3 TASK DEFINITION

In this section, the proposed system is described briefly along with details about the corpus development for SA of Bangla News at the document level.

16.3.1 Identifying Sentiment from Bangla news documents

Our proposed system aims to identify sentiments in the Bangla news as positive or negative.

16.3.1.1 Positive News (PN)

News content that is encouraging, uplifting, desirable and positively impacts society is considered as positive news.

16.3.1.2 Negative News (NN)

News content that contain any kind of unfavorable information and negatively impacts society is considered as negative news.

16.3.2 Corpus Development

In this section, the whole process of data preparation before processing is discussed.

16.3.2.1 Data Collection

In this era of technology, online news is becoming one of the largest sources of news. We collected our Bangla News for SA tasks from online Bangla news portals as that is the most reliable source of news. News content of different categories were collected using web scraping and stored for further use. All the news content were collected from popular news sites of Bangladesh such as prothomalo.com, bd-news24.com, bbcbanglanews.com and kalerkontho.com.

16.3.2.2 Data Pre-processing

Text documents are written following the linguistic characteristics with particular arrangement of characters, numbers, words and expressions.

Before the data is fed to the classifier, it must be prepared so that the experimental results meet the desired expectations. To represent data as a vector space so that it can be fed to classifiers and to remove non-informative features, certain preprocessing steps were taken such as Tokenization, Removal of Punctuation Words, Digit Removal, and Stop Words Removal. After preprocessing, the data is clean, which can lead to better outcomes.

16.3.2.3 Data Annotation

Four different annotators were given the collected Bangla News to manually label of Positive News (PN) or Negative News (NN). Some lists of conditions were given to all of them that were selected and approved by themselves. So, with mutual approval of the four annotators, the conditions that are needed to annotate the news were approved. Disagreement among the annotators was discussed and voting was applied to give the news a label of positive news or negative news. Data statistics are shown in Table 16.2.

16.4 METHODOLOGY

In this section, all the methods used in our proposed system are described briefly. The whole procedure is shown in Figure 16.1. After accumulating the pre-processed news documents, different feature extraction techniques are implemented before fitting into the supervised ML classifiers and LSTM model. In the next phase, six ML classifiers SVM, DT, MNB, KNN, RF and LR are employed individually to build a predictive model.

Similarly, a deep learning-based approach LSTM is implemented to analyze the textual pattern and extract sentiments from individual documents. In this way, results from both approaches are combined and a comparative analysis of the models is estimated. In the following subsections, different parameters and techniques involved in the proposed system are discussed.

16.4.1 Feature Extraction

Both of the supervised ML approaches and the LSTM method can perform efficiently by extracting features before training it with the dataset. Consequently, in this system, two different techniques are used to accomplish this task. In terms of supervised ML model, unigram features have been extracted using the Count Vectorizer and TF-IDF techniques. Through tokenization of all the text documents, a matrix of word vectors is formed using the Count Vectorizer technique. The formed matrix is

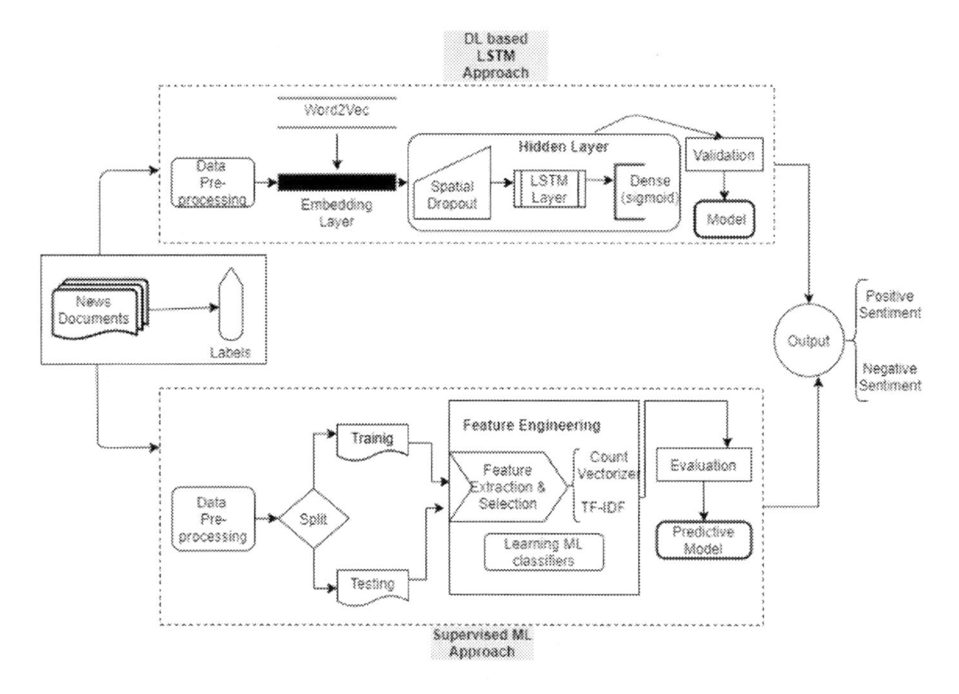

Figure 16.1 Architecture of Proposed Model for SA of Bangla news

processed through the TF-IDF technique, which removes the incorporated biasness due to repetitive word frequencies and large texts. Considering the DL based approach LSTM, the Word2Vec embedding technique is used to extract features. This process helps transform the texts into dense vectors by interpreting the semantic structure and is accessed through the Keras library. In our system, the embedding dimension is tuned to 100 with kernel size 3, batch, and hidden dimension size of 250.

16.4.2 Supervised ML algorithms

As the supervised ML approach is commonly used in SA studies, in our system six such popular approaches are used. SVM, DT, MNB, KNN, RF and LR are individually implemented to analyze the behavior of the dataset in each case. The reasons behind adopting these classifiers are discussed below briefly.

MNB is such an approach that considers conditionality of texts and also the case of word frequencies. This approach is mostly applied in case of classifying texts [8][9].

In this work, Linear SVM is employed with alpha parameters having the value of 1e-3 with a random state of 40 and constant learning rate. It has been seen from various previous studies that SVM shows promising results for classification tasks due to its optimal hyperplane feature [10].

The LR classifier basically performs well for textual data due to its ability to extract weights from the features and predict directions by separating observations independently [11].

KNN is useful in making faster predictions. This classifier can handle multi-classification problem[12]. For our system, n-neighbors with parameter 20 is implemented.

RF is employed in this work with maximum features of 50 having the total value of number of trees as 300. The ability of this classifier to perform effectively in a categorical domain is the major factor in its selection. Moreover, this classifier builds different decision tress independently and selects the best model through voting [13].

The DT classifier estimates the best possible data partition by calculating information gain. Moreover, as textual data includes outlier data, this kind of classifier gives the best possible outcome in such cases [14].

16.4.3 Deep learning approach LSTM

LSTM propagates information through sequential pattern, feeding data of the immediate past state information and linking it with the next one [15]. To tune the parameters for LSTM, this model has optimally chosen the values. In this case, the dropout rate is taken as 0.5 with 2 dense layers. Evaluating the model with the "sigmoid" activation function and "binary cross entropy" loss function, the model is executed 100 epochs with a default batch size of 32.

16.5 EXPERIMENTS AND RESULT ANALYSIS

In this section, experimentation procedure and insights of results are discussed.

16.5.1 Performance Measurement Tools

To identify the sentiment of Bangla news as Positive News (PN) or Negative News (NN) we developed this model. We calculated the precision, recall and f1 score of each class to evaluate the model. Six ML classifiers, SVM, DT, MNB, KNN, RF, LR, and one deep learning method, LSTM, are employed to build the model. Google Collaboratory platform is used to implement the model. We used a percentage split method to split the dataset for training and testing purposes of the model. Here 80% of the data is used for training and 20% is used for testing. In the case of LSTM, testing data is further split into validation data. Here, 20% of the testing data is considered as validation data. Random shuffling was done for better results of the model.

16.5.2 Experimental Output

Some of the sample inputs, actual value, and its prediction by our model are portrayed in Table 16.3.

16.5.3 Performance Statistics

Results of six ML classifiers, RF, LR, DT, KNN, MNB, SVM and one deep learning method LSTM on our model is evaluated by Precision, Recall, F1-Score and Weighted Average on each case. Table 16.4 shows the classification report of six ML classifiers in detail along with 10-fold cross validation. Table 16.5 shows the classification report

Table 16.3 Sample of a few predicted output results for news documents

Input Data	Actual Data	Predicted Value
"করোনাভাইরাসের টিকা নেওয়ার ক্ষেত্রে অনেক দেশেই কিছু মানুষের অনাগ্রহ দেখা যাচ্ছে। টিকা কতটুকু নিরাপদ হবে বা এর পার্শ্বপ্রতিক্রিয়া কেমন হবে, এ নিয়ে মানুষের মধ্যে সংশয় আছে। এ ছাড়া টিকা নিয়ে আছে নানা অসত্য তথ্য বা গুজব। ফলে মানুষকে টিকা দিতে উৎসাহী করে তুলতে অভিনব পন্থা নিয়েছে থাইল্যান্ডের কর্তৃপক্ষ।" "Some people in many countries are reluctant to get vaccinated against coronavirus. People are skeptical about how safe the vaccine will be or what the side effects will be. There are also misinformation or rumors about the vaccine. The Thai authorities have taken over".	PN	PN
"ছয় জেলায় করোনার টিকা ফুরিয়ে গেছে। রোববার এসব জেলায় কোনো ব্যক্তিকে টিকা দেওয়া হয়নি। এসব জেলায় আবার কবে টিকা দেওয়া হবে, তা জানেন না স্থানীয় স্বাস্থ্য বিভাগের কর্মকর্তারা।গাজীপুর, রাঙামাটি, রাজশাহী, পাবনা, নড়াইল ও চুয়াডাঙ্গা জেলায় করোনার দ্বিতীয় ডোজ দেওয়ার মতো টিকা নেই। গতকাল রাজধানীর মুগদা মেডিকেল কলেজ হাসপাতালেও কোনো টিকা দেওয়া হয়নি। স্বাস্থ্য অধিদপ্তর থেকে দেওয়া পরিসংখ্যানে করোনা টিকা পরিস্থিতি সম্পর্কে এ তথ্য পাওয়া গেছে।" "Corona vaccination has run out in six districts. No one was vaccinated in these districts on Sunday. Local health officials do not know when they will be vaccinated again in these districts. No vaccine. No vaccine was given at Mugda Medical College Hospital in the capital yesterday. Statistics from the Department of Health provide information about the corona vaccine situation".	NN	NN

of the deep learning method LSTM in detail. Pictorial visualization with six ML classifiers in both evaluation criteria, i.e., percentage split method and 10-fold cross validation method can be found from Figure 16.2. It shows minimal distance between both criteria, which is an indication of good fitting of the model.

Both Figure 16.2 and Table 16.3 show that the RF classifier achieved the highest accuracy of 80% among other ML classifiers. But after comparing with the deep learning method LSTM, we can see that LSTM achieved the highest accuracy of 83%.

Ranking order of the classifier from high to low performance is LSTM, RF, DT, LR, SVM, KNN and MNB. So MNB shows the lowest performance with 70% accuracy, which is also quite good behavior. So, the deep learning method LSTM is showing better results than the ML classifiers.

Figure 16.3 shows graphical representation of ROC curve analysis of the six supervised ML classifiers. Analyzing the curve, it can be seen that all the classifiers have shown good performance in terms of having higher true positive rates. However, SVM, RF and MNB have shown comparatively quite close efficiency. On the other hand, the KNN classifier has shown the lowest efficiency in comparison to others.

Table 16.4 Classification report of six supervised classifiers for SA using the ML approach in both split and 10-fold cross validation technique.

Evaluation Criteria	Type	Measures	MNB	SVM	LR	KNN	RF	DT
Split Method	Negative	P	0.7	0.77	0.74	0.75	0.78	0.81
		R	1.0	0.89	0.75	0.89	0.97	0.90
		F1 score	0.82	0.83	0.73	0.82	0.87	0.85
Split Method	Positive	P	1.0	0.66	0.68	0.63	0.87	0.73
		R	0.08	0.45	0.48	0.38	0.44	0.55
		F1 score	0.15	0.54	0.56	0.48	0.58	0.63
Split Method	Weighted	P	0.79	0.74	0.73	0.71	0.81	0.78
		R	0.70	0.75	0.69	0.73	0.80	0.79
		F1 score	0.61	0.73	0.70	0.71	0.78	0.78
Split Method		Total Accuracy	0.70	0.75	0.76	0.73	0.80	0.79
10-fold cross validation		Accuracy	0.69	0.76	0.78	0.71	0.82	0.85

Table 16.5 Classification Report of DL based approach LSTM in SA.

DL Approach	Negative			Positive			Weighted			Total Accuracy
	P	R	F1 score	P	R	F1 score	P	R	F1 score	
LSTM	0.86	0.89	0.88	0.76	0.70	0.73	0.83	0.83	0.83	0.83

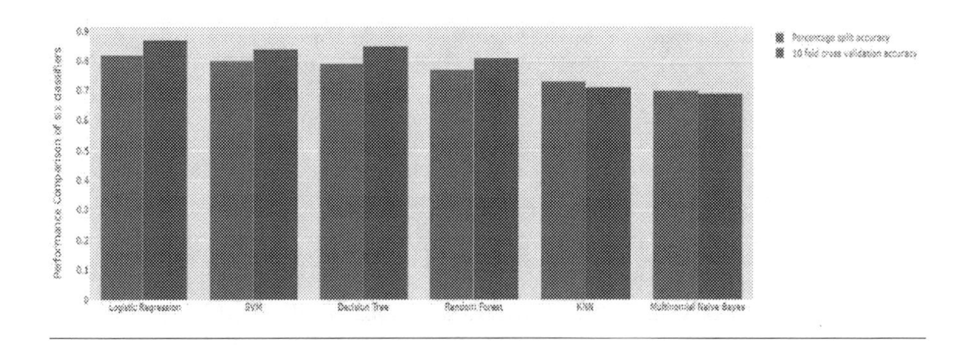

Figure 16.2 Performance analysis of classifiers in bar chart

Thus, it can be understood from the overall analysis that RF has performed well in comparison to all the six supervised ML approaches adopted in this system. Consequently, LSTM outperformed all other models with promising prediction outcomes..

16.5.4 Error Analysis

While testing our model with different sample input, we observed some misclassification made by our model. After testing the model with some selected critical sample input, we tried to understand the reasons behind this behavior of the model. According to our observations, scarcity of positive news in comparison to negative news can

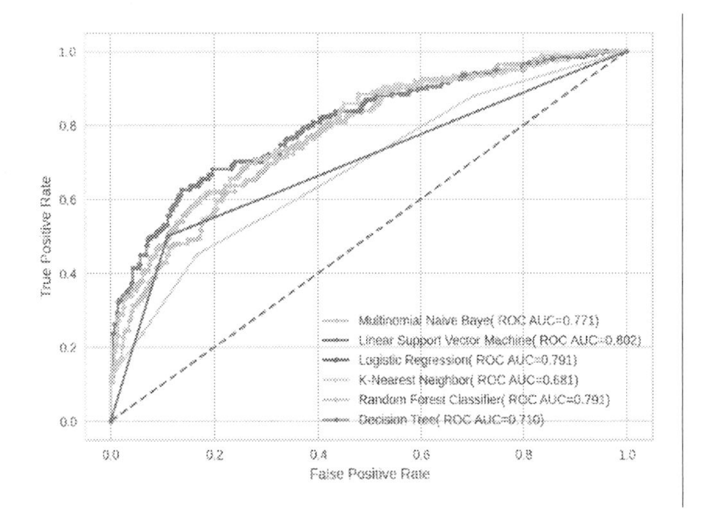

Figure 16.3 ROC curve performance of six supervised ML classifiers

Table 16.6 Error analysis for critical data

Input Data	Actual Data	Predicted Value
"বাংলাদেশ প্রকৌশল বিশ্ববিদ্যালয়ের (বুয়েট) ছাত্র আবরার ফাহাদ হত্যা মামলায় তাঁর রুমমেট মিজানুর রহমানকে কারাগারে পাঠানোর আদেশ দিয়েছেন আদালত। পুলিশের আবেদনের পরিপ্রেক্ষিতে ঢাকার মুখ্য মহানগর হাকিম আদালত আজ মঙ্গলবার এ আদেশ দেন।" "A court has ordered to send his roommate Mizanur Rahman to jail in connection with the murder of Bangladesh University of Engineering and Technology (BUET) student Abrar Fahad. The Dhaka Chief Metropolitan Magistrate's Court passed the order on Tuesday following a police appeal".	Neutral News	NN

be one problem behind this misclassification. So, we have a future plan to increase the number of positive news content in our dataset. Another reason is not considering the neutral news content which has led to difficulties in processing. Table 16.6 shows an example of a misclassification of the model where neutral news is given as sample input and the model classified it as Negative News (NN).

16.6 CONCLUSION

The Sentiment Analysis (SA) of Bangla News where six state-of-the-art supervised machine learning algorithms, including DT, MNB, KNN, LR, RF, SVM and one deep learning method LSTM are used to choose the best among them for our domain. The LSTM algorithm outshines all other algorithms securing 83% accuracy. MNB got the lowest accuracy of 70%. In this work, there have been many challenges i.e., unbalanced data to work with, changes in behavior of the algorithms with dataset, difficulty in hyperparameter settings, etc.

The main future direction of our research is to add neutral polarity as neutral news is mistreated, enlarge our dataset efficiently by adding content with mixed sentiments, analyze semi-supervised learning and unsupervised learning approaches, evaluate other deep learning algorithms for more efficient SA and for more effective comparison.

Bibliography

[1] Devika, MD and Sunitha, and Ganesh, Amal(2016). Sentiment analysis: A comparative study on different approaches. *Procedia Computer Science*, 87, pp. 44–49, Elsevier.

[2] Chowdhury, Shaika and Chowdhury and Wasifa(2014). Performing sentiment analysis in Bangla microblog posts. *2014 International Conference on Informatics, Electronics & Vision (ICIEV)*, pp. 1–6, IEEE.

[3] Hassan, Asif and Amin, Mohammad Rashedul and Mohammed, N and Azad, AKA(2016). Sentiment analysis on Bangla and romanized Bangla text (BRBT) using deep recurrent models.*arXiv preprint arXiv:1610.00369.*

[4] Tabassum, Nusrath and Khan and Muhammad Ibrahim(2019). Design an empirical framework for sentiment analysis from Bangla text using machine learning.*2019 International Conference on Electrical, Computer and Communication Engineering (ECCE)*, pp. 1–5, IEEE.

[5] Wahid, Md Ferdous and Hasan, Md Jahid and Alom, Md Shahin(2019). Cricket sentiment analysis from Bangla text using recurrent neural network with long short term memory model.*2019 International Conference on Bangla Speech and Language Processing (ICBSLP)*, pp. 1–4, IEEE.

[6] Arid Hasan, Md and Tajrin, Jannatul and Absar Chowdhury, Shammur and Alam, Firoj(2020). Sentiment classification in Bangla textual content: A comparative study.*arXiv e-prints*, pp. arXiv–2011.

[7] Hussain, Md Gulzar and Hasan, Md Rashidul and Rahman, Mahmuda and Protim, Joy and Al Hasan and Sakib(2020). Detection of Bangla fake news using MNB and SVM classifier.*2020 International Conference on Computing, Electronics & Communications Engineering (iCCECE)*, pp. 81–85, IEEE.

[8] Abbas, Muhammad and Memon, K Ali and Jamali, A Aleem and Memon, Saleemullah and Ahmed and Anees(2019). Multinomial Naive Bayes classification model for sentiment analysis.*IJCSNS Int. J. Comput. Sci. Netw. Secur*, 19(3), pp. 62–67.

[9] Sharif, Omar and Hoque, M. M. and Hossain, Eftekhar(2019). Sentiment analysis of Bengali texts on online restaurant reviews using multinomial Naive Bayes.*2019*

1st International Conference on Advances in Science, Engineering and Robotics Technology (ICASERT), (pp. 1–6). IEEE.

[10] Rahat, Abdul Mohaimin and Kahir, Abdul and Masum, Abu Kaisar Mohammad(2019). Comparison of Naive Bayes and SVM Algorithm based on sentiment analysis using review dataset.*2019 8th International Conference System Modeling and Advancement in Research Trends (SMART)*, pp. 266–270, IEEE.

[11] Al-Amrani, Yassine and Lazaar, Mohamed and Elkadiri and Kamal Eddine(2017). Sentiment analysis using supervised classification algorithms.*Proceedings of the 2nd International Conference on Big Data, Cloud and Applications*, pp. 1–8.

[12] Huq, Mohammad Rezwanul and Ali, Ahmad and Rahman and Anika(2017). Sentiment analysis on Twitter data using KNN and SVM.*International Journal of Advanced Computer Science and Applications*, 8(6), pp. 19–25.

[13] Karthika, P., Murugeswari, R., and Manoranjithem, R. (2019). Sentiment analysis of social media network using random forest algorithm.*2019 IEEE International Conference on Intelligent Techniques in Control, Optimization and Signal Processing (INCOS)*, pp. 1–5, IEEE.

[14] Kim, Han-Min and Park and Kyungbo(2019). Sentiment analysis of online food product review using ensemble technique.*Journal of Digital Convergence*, 17(4), pp. 115–122, The Society of Digital Policy and Management.

[15] Luan, Yuandong and Lin, Shaofu(2019). Research on text classification based on CNN and LSTM.*2019 IEEE International Conference on Artificial Intelligence and Computer Applications (ICAICA)*, (pp. 352–355). IEEE.

CHAPTER 17

Employee Turnover Prediction Using a Machine Learning Approach

Md. Ali Akbar

Department of Computer Science & Engineering, East Delta University, Chattogram, Bangladesh

Kamruzzaman Chowdhury

Department of Computer Science & Engineering, East Delta University, Chattogram, Bangladesh

Mohammed Nazim Uddin

School of Science, Engineering & Technology, East Delta University, Chattogram, Bangladesh

CONTENTS

EMployee turnover is identified as a major issue for any business leaders due to its adverse impact on workplace productivity and affect on organizational expenses. Loss of a talented employee is a loss of tactical information which may bring competitive benefit to another organization. Minimizing employee attrition is a key challenge in achieving the end goal and staying in the market. The aim of this chapter is to prepare a machine learning model that can predict employees who will leave the organization and determine the characteristics of both active and terminated employees dependent on HR analytics dataset. Accurate prediction enables an organization to discover the reasons why workers leave and help it take early action to retain employees or prepare for succession planning.

17.1 INTRODUCTION

Staff retention is very important for any companies in sustaining their business. A high turnover rate is a huge organizational expense. Expenditures may include the costs of recruiting, selecting and training possible replacements. It affects overall productivity and ongoing work of existing employees. The reason is that the new employee takes time to settle into the new environment before they can be effectively productive. Sometimes employees leave even though they have good performance and a good working environment.

One way to reduce the turnover rate is to predict employee attrition, which will help the management upgrade their internal policies. A machine learning algorithm can help the organization for predict the employee attrition rate. Feature analysis of historical data kept in HR departments and preparing a machine learning model can help to determine the characteristics of employees who will leave the organization.

We have divided our research into 5 parts or sections. Section 17.2 is the literature review. It offers an outline of common understanding, which allows you, in ongoing research, to identify ideas, approaches and gaps. It is basically the work related to previous research. In Section 17.3, we introduce our system architecture. In Section 17.4, we describe the experimental evaluation. Finally, in Section 17.5, we demonstrate the conclusion of our research.

17.2 RELATED WORKS

Employee turnover is a high impact topic and has been researched in both management as well as psychology studies. Employee turnover can be defined as "the gross movement of workers in and out of employment with respect to a given company" [1]. According to the Bureau of Labor Statistics, there are two categories of turnover. Employee turnover is measured in terms of persons that quit their job, also known as voluntary turnover, and total separations. Subtracting the voluntary turnover from the total separations will give you the amount of involuntary turnover. This research focuses on voluntary turnover and the reason behind it. Research suggests four categories of factors when it comes to employee turnover. Organization-wide factors, immediate work environment factors, job-related factors, and personal factors [2]. In [3], the authors indicate in their model a multi-layered approach to turnover motivators including traditional features such as job satisfaction, meeting of expectations, or job involvement, as well as newer attitudes like stress, psychological uncertainty, challenge, hindrance stressors, or organizational context like company size, group cohesion, and demography. General job availability, movement of capital, and job satisfaction can interact with each other simultaneously to affect turnover [4]. The relationship between turnover intentions and turnover, for example, can be moderated by various personality traits. This relationship was found to be stronger for employees with low self-monitoring, low risk aversion, and an internal locus of control [5]. Some research explored the relationship between job satisfaction and turnover and found it to be significant and consistent, but not particularly strong [6], Capturing all these dimensions has a problem. A more complete understanding of the psychology of the withdrawal decision process requires investigation beyond the replication of the satisfaction turnover relationship. Reference [7] recommends a thorough comparison of several models and data exploration. Explicitly stating the need to continue model testing rather than simply correlating variables with turnover. In [8], the authors found that most studies of voluntary turnover have one or two independent variables with voluntary turnover as the dependent variable. Most of the analytic reviews of voluntary turnover found that the strongest predictors for voluntary turnover were job satisfaction, salary, marital status, age, working condition, growth potential, etc. [9, 10, 11]. Among five ML algorithms and three feature selection techniques, the best predictors were identified using the selectKBest, Recursive Feature Elimination and Random Forest models to predict whether employees would leave a company [12]. Different ML algorithms were trained which included a support vector machine, logistic regression, Adaboost with optimal hyperparameters, Recursive Feature Elimination, Native Bayes, and Decision Tree. Among them, two predictive models have better performance: the SVM-polynomial kernel using RF and DT with Select KBest [13]. Through a comprehensive and robust evaluation process, the performance of various supervised machine learning approachs are analyzed and reliable guidelines are provided [14]. This study shows that because of the regularization formulation, the Extreme Gradient Boosting technique is more robust, and six historically used supervised classifiers were compared with XGBoost [15].

17.3 METHODOLOGY

This section sheds light on the methodology of the proposed model.

17.3.1 System Architecture.

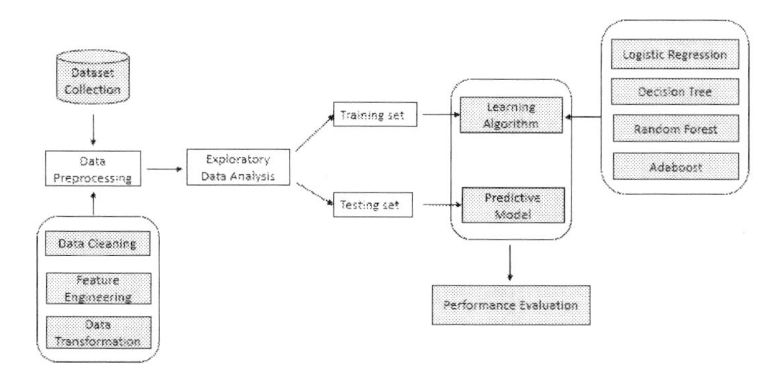

Figure 17.1 Proposed System Architecture.

17.3.2 Dataset Collection

For this project, an HR dataset named HR_comma_sep has been picked, which is from Kaggle. This data set contains 14999 records of employees with 10 features satisfaction-level, last-evaluation, number-project, average-monthly-hour, time-spend-company, work-accident, left, promotion-last-5 years, department, and salary.

17.3.3 Data Preprocessing

There are several steps in data preprocessing for this proposed model.

17.3.3.1 Data Cleaning

There are many opportunities for a dataset to be mislabeled or duplicated when combining multiple data sources. In order to get reliable outcomes from the algorithm, the dataset needs to be cleaned. This dataset is super cleaned and no missing values were found.

17.3.3.2 One Hot Encoding

While working with real-world data, it is difficult to fit a machine learning model into the dataset because the dataset may contain numeric as well as categorical data. Some algorithms can learn directly from categorical data like the decision tree algorithm, but many algorithms can learn only from numerical data. The dataset that is used in this research contains two categorical variables: department and salary. In order to convert these categorical variables into the numerical variables, the One Hot Encoding technique is used.

17.3.3.3 Feature Selection

Recursive Feature Elimination (RFE) operates by recursively eliminating variables and creating a model for the remaining variables. After applying one hot encoding, we got 20 features. From these features, we select the top 10 features that most affects employee attrition through feature selection.

17.3.3.4 Dataset Split

For training and testing of data, the train test split function is necessary to split the data collection into two subsets. The training set is used to train the model and the test set is a subset to test the model. In this study, 80% of the data is used as a training set and the remaining 20% of the data is used as a test set.

17.3.4 Class Imbalance

We've got an imbalanced dataset. In addition, class imbalance affects an education algorithm by indirectly rendering the majority-class judgment rule a formula that optimizes predictions based on a majority class in the dataset. This topic can be answered in three ways:

a. Give the minority class a bigger penalty for false predictions.

b. Up-sampling the class of minorities or down-sampling the class of majority.

c. Generate examples of synthetic training.

Here, we use up-sampling by a library function called "from imblearn.over_sampling import SMOTE". This object is a SMOTE implementation - Synthetic Minority Over-sampling Technique.

17.3.5 Performance Matrix

For any classifier, accuracy is an important metric. But we should also consider true positives (TP), false positives (FP), true negatives (TN), false negatives (FN), recall, and precision. Since the cost associated with false positives and false negatives is different, these metrics become necessary.

17.3.6 Selected Classification Methods

Eventually, we used four different algorithms for prediction. This section relates to the concept behind different supervised classification algorithms.

17.3.6.1 Base Rate Model

Here we calculate a base rate model that always chooses the majority class of the target variable. It is only used to assess how much better that one would be. In this dataset, the expected majority class is 0 and it is workers who have not left the company. When you remember: Data exploration, 24% of the dataset contains 1s (the worker who quit the company) and the rest 76% contains 0s (workers who did not quit the company). It just forecasts every 0s and excludes all 1s. Example: For this data set, the base rate accuracy of all data must be 76%, since 76% of the data set is classified as 0 (employees not leaving the company).

17.3.6.2 Logistic Regression Classifier

One of the fundamental linear classification models is logistic regression (LR). It is ideally suited for forecasting binary or non-dependent variables. Additional benefits are that the model and the correlation between characteristics offer many ways to regularize. Another benefit is that it is easy to add new data and update the model in a continuous process. This model is widely used to estimate the probability that there may be an occurrence. The output is either 0 or 1. This is a valuable model for our situation, since we are interested in predicting whether an employee will leave (0) or stay (1). Logistic regression forecasts the outcome of the response variable by a number of other explanatory variables, often referred to as predictors. In this domain, the value of our answer variable is classified into two forms: 0 (zero) or 1 (one). The value of 0 (zero) is the likelihood that the employee will not leave the company and the value of 1 (one) is the probability that the employee will leave the company.

17.3.6.3 Decision Tree Classifier

Decision Tree (DT) model is another feasible alternative. Decision Trees may be inconvenient if you require a precise rotational forecast model that uses fresh data and updated trends each year. A decision tree follows a tree structure in which each node is an attribute test. Each branch in the tree is the product of the test, and each leaf node is a class label. Decision nodes, chance nodes and end nodes can be located in the decision tree.

17.3.6.4 Random Forest Classifier

Ensemble approaches include trees like the Random Forest (RF). The capacity of RF to embrace non-linear functionality could prove itself. This will be critical for potential implementations as the dataset expands. However, its lack of exposure to associated functionality is a big drawback of RF models. Strong features may end up with low values with associated functions.

17.3.6.5 AdaBoost Classifier

Adaptive Boosting is a boosting learning algorithm which combines simple models (weak learners) in order to build a strong learner sensitive to noisy data and outliers and less prone to overfitting. It is used to reduce the variance and bias for supervised learning.

17.4 EXPERIMENTAL EVALUATION

Statistical results are included in this section.

17.4.1 Exploratory Data Analysis

The heat map in Figure 17.2 shows that there is a clear positive association among project count, average monthly hours, and evaluation. There is a 35% association

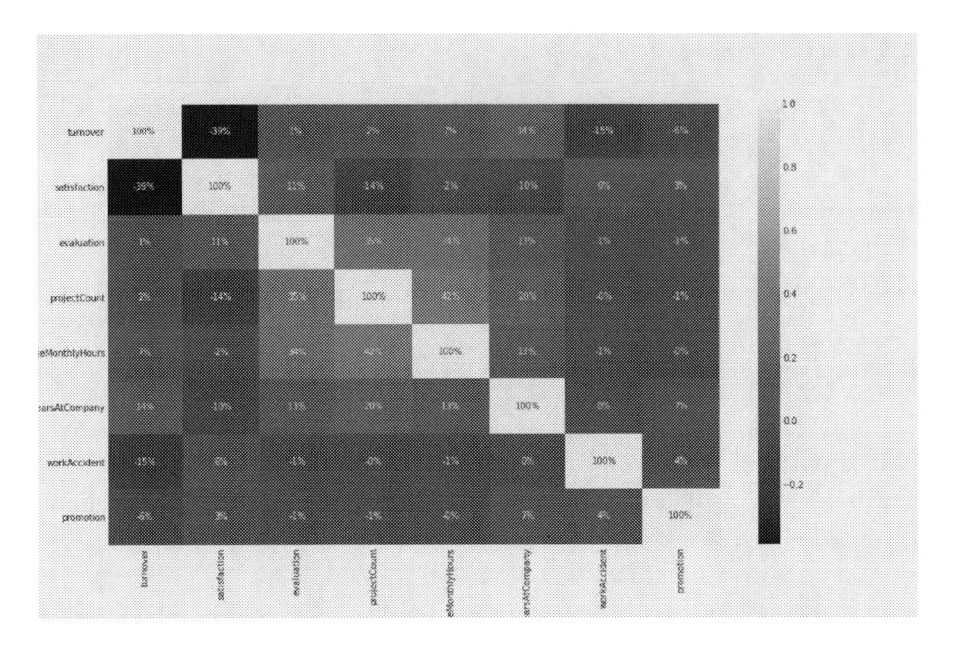

Figure 17.2 Correlation Matrix & Heat map.

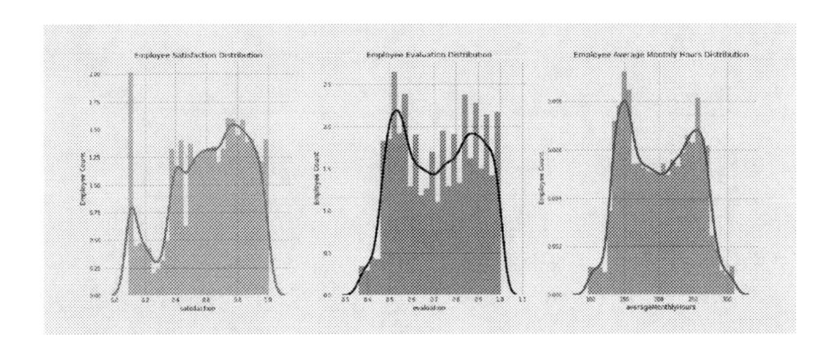

Figure 17.3 Satisfaction vs Evaluation vs Average Monthly Hours.

between project count and evaluation, 42% between project count and average monthly hours, 34% between average monthly hours and evaluation, and on the other hand, a moderate negative correlation between satisfaction and turnover, which is - 39%.

Figure 17.3 shows that there is a similar bimodal distribution in both evaluation and average monthly hours means that employees with high average monthly hours were evaluated highly and vice versa, which is already indicated in the high positive correlation between evaluation and average monthly hours in the correlation matrix. The employee satisfaction distribution plot shows that there is a huge spike for workers with low satisfaction and high satisfaction.

Employees with average and lower salaries tend to leave the organization more and a small number of employees left with high salaries.

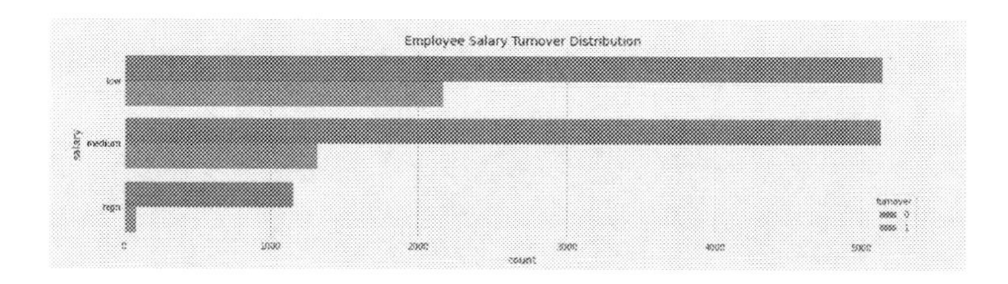

Figure 17.4 Salary vs Turnover.

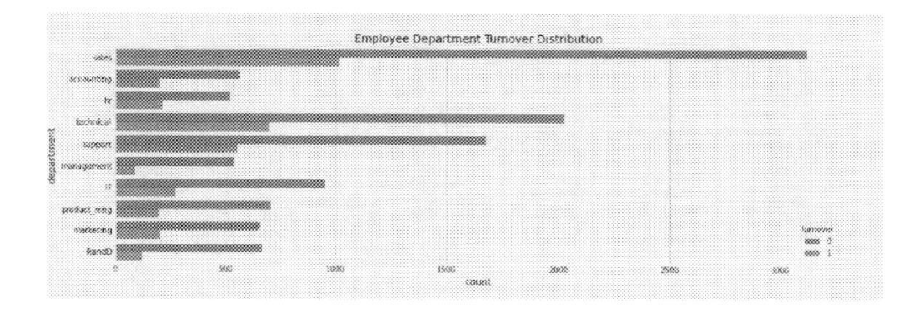

Figure 17.5 Department vs Turnover.

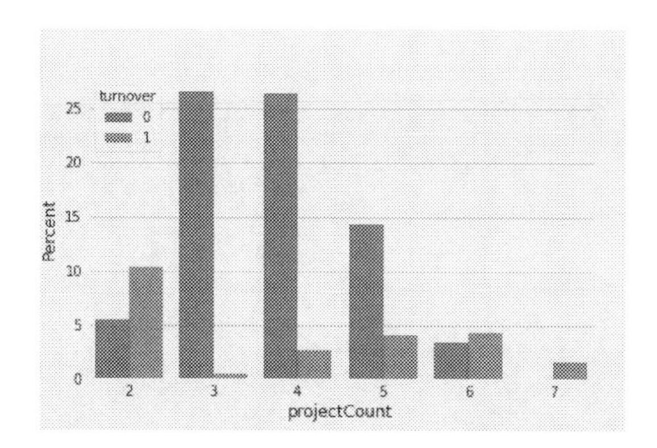

Figure 17.6 Project count vs Turnover.

Let's look at an employee department vs turnover plot, which shows the sales, technical and support departments have a high turnover rate and the management department had less turnover.

When the analysis was done for the number of projects, more than half of employees with 2, 6, and 7 projects left the organization and all employees with 7 projects left, but most employees with 3, 4, and 5 projects did not leave, and turnover of employees with 3 projects was very low.

From the above graph points, most of the employees left after working for 3, 4, and 5 years. More than half of the employees left at 5 years. After 6 years there is no turnover.

The bi-modular distribution of turnover vs. evaluation indicates that employees with higher (.8-1) evaluation scores and lower (.4-.6) evaluation scores tend to leave the organization more. Evaluation scores between .6-.8 indicate lower levels of turnover.

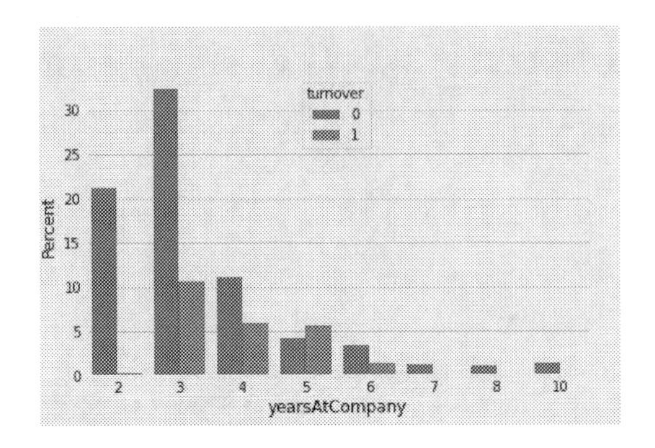

Figure 17.7 Turnover vs Years at Company.

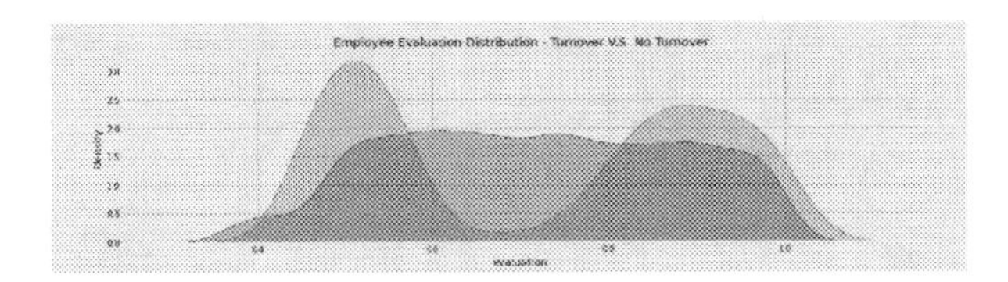

Figure 17.8 Turnover vs Evaluation.

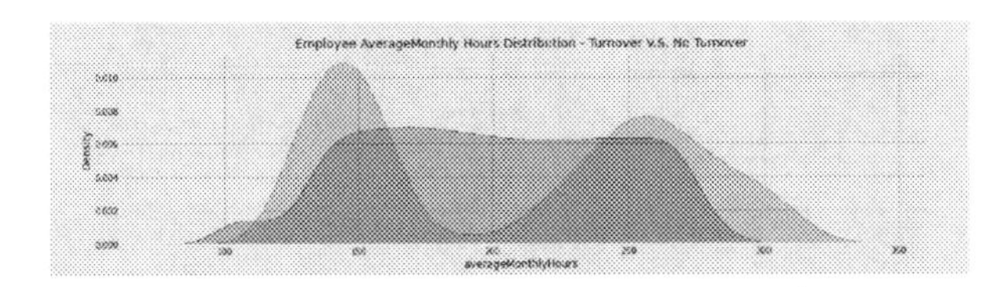

Figure 17.9 Turnover vs Average Monthly Hours.

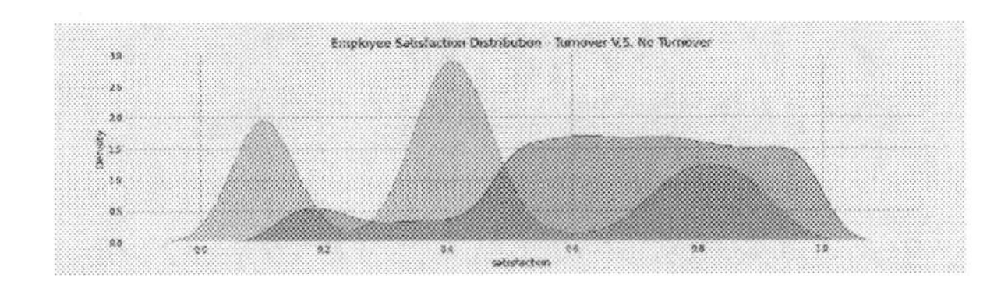

Figure 17.10 Turnover VS Satisfaction.

Another bi-modular distribution of average monthly hours indicates that employees with minimum and maximum hours of work left the organization, which means they are underworked or overworked.

A tri-modal distribution of employees that had the turnover in Figure 17.10 indicates that employees with low satisfaction left the company.

The box plot shown in Figure 17.11 indicates the average monthly hours increase along with the number of projects for employees who left the company. At the same time, those who had consistent monthly hours with an increasing number of projects didn't leave the organization. Employees who did leave needed much time to complete their work.

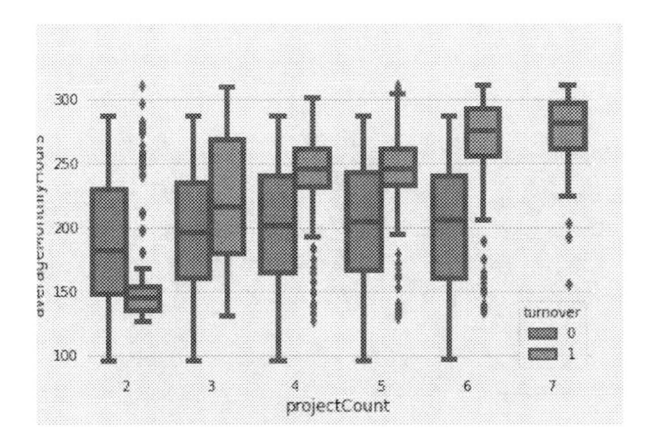

Figure 17.11 Project Count vs Average Monthly Hours.

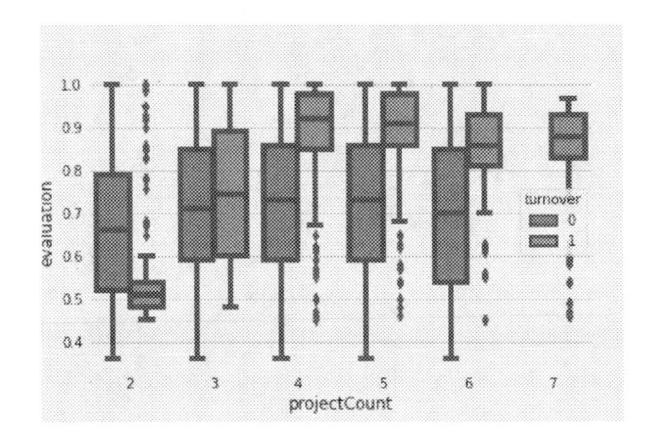

Figure 17.12 Project Count vs Evaluation.

Similarly, the box plot between project count and evaluation showed results very similar to the graph above but the evaluation score is higher along with the increasing number of projects in the turnover group, so they might have better job opportunities.

Analysis of the graph in Figure 17.13 indicates three clusters of employees who left the organization, they are:

Cluster 1: Employees with higher evaluation scores but a lower satisfaction level are hard workers but feel unhappy in the organization. They might be overworked.
Cluster 2: The employees with a lower satisfaction level (.35-.45) and lower evaluation scores (less than .6) are badly evaluated and also feel unhappy. They might be underworked.
Cluster 3: Employees with higher evaluation scores and a higher satisfaction level could be hard working and happy employees. They are the ideal employees of the organization, but must have left the company because of a better job offer.

Figure 17.14 indicates three clusters of employees leaving the organization based on their satisfaction & evaluation:
Cluster 1 (Blue): Good employee but felt horrible at work.
Cluster 2 (Green): Felt horrible at work and badly evaluated.
Cluster 3 (Red): Great working performance but left due to better job opportunities.

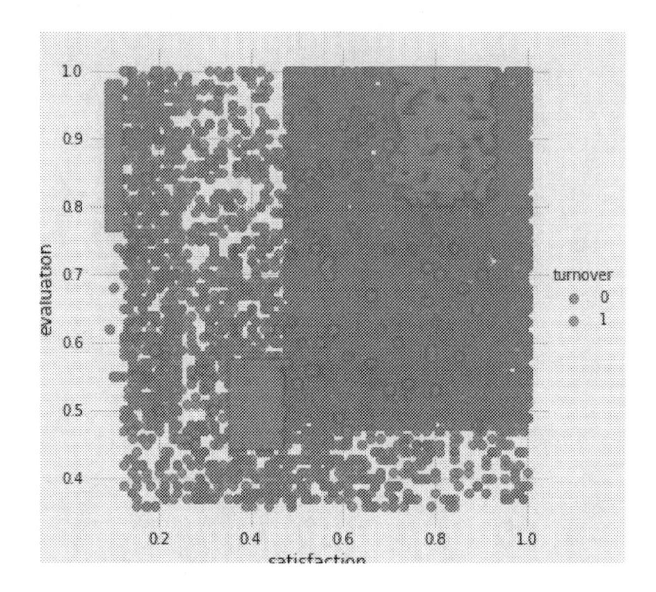

Figure 17.13 Satisfaction VS Evaluation.

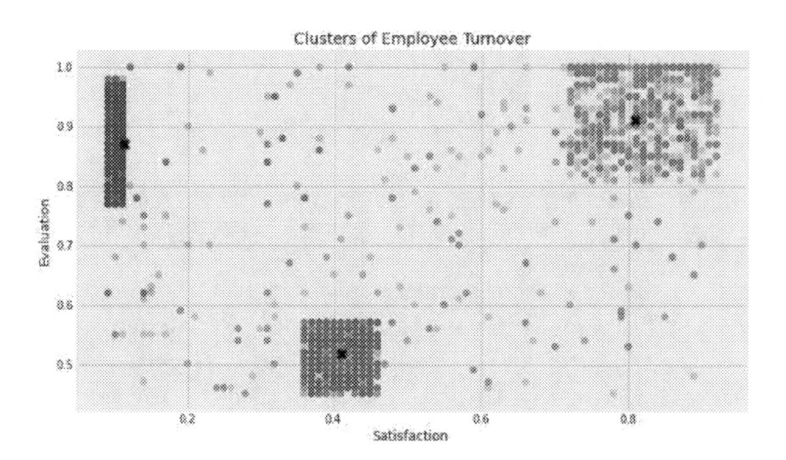

Figure 17.14 K-Means Clustering of Employee Turnover (satisfaction vs. evaluation).

Table 17.1 Results for Different Classification Methods

Method	Accuracy	Precision	Recall	F1 Score	CV Score
Logistic Regression	76%	50%	77.45%	60.76%	81.6%
Decision Tree	96%	89.2%	93.69%	91.39%	96.3%
Random Forest	98%	94.57%	95.23%	94.90%	99.3%
Ada Boost	94%	81.32%	94.53%	87.43%	96.7%

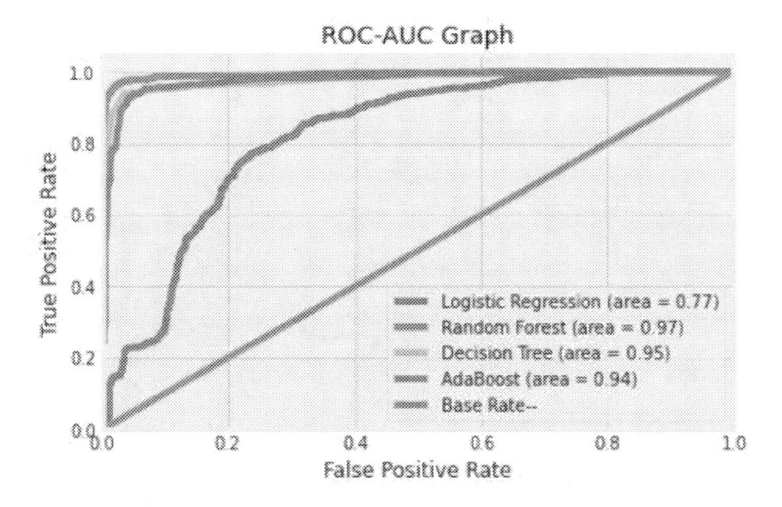

Figure 17.15 ROC-AUC Graph.

Table 17.2 AUC Score.

Learning Algorithm	AUC Score
Logistic Regression	0.76
Decision Tree	0.95
Random Forest	0.97
Ada Boost	0.94

17.4.2 Result Analysis

17.4.2.1 ROC-AUC Graph

The ROC - AUC curve is a classification challenge efficiency calculation at different threshold conditions. The receiver operating characteristic curve (ROC) is measured using true positive rate and false positive rate coordinates. The ROC and AUC scores for the different machine learning algorithms used in this research are shown in Figure 17.15. The ROC graph indicates that the Random Forest algorithm achieved the highest AUC score, which is 0.97.

17.4.2.2 Feature Importance

Figure 17.16 shows the most important features, according to the Random Forest algorithm, that influence when an employee exits the organization in ascending order.

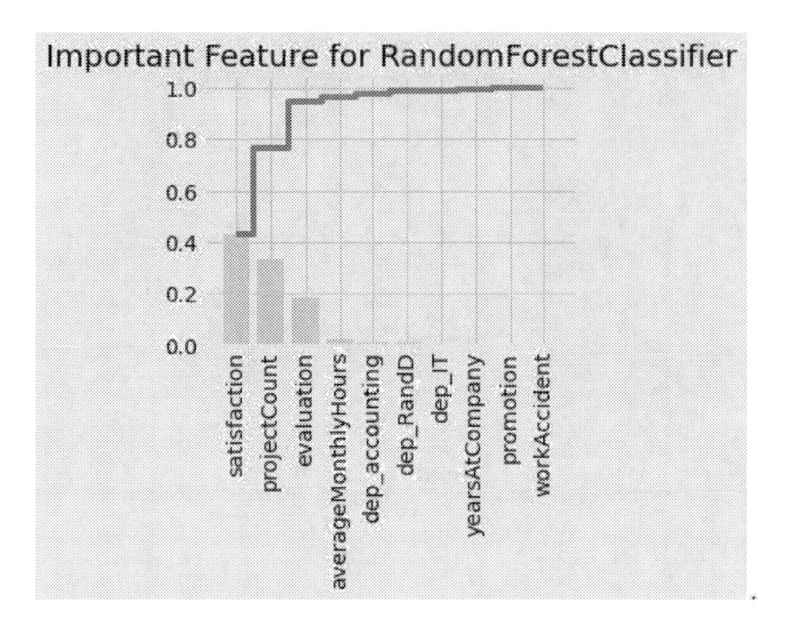

Figure 17.16 Feature Importance by Random Forest Classifier.

17.5 CONCLUSION

In this chapter, we concentrate on the employee attrition problem to see how the employee leaves his organization after a given amount of time. We articulate this as a problem of binary classification that categorizes workers as those who are interested in leaving the company. Our approach is focused primarily on the use of machine learning methods. Our findings in this analysis suggest that the current company roles are listed as the appropriate attributes after the feature selection process. To get comparative results, Logistic Regression, Decision Tree, Random Forest, and Ada Boost algorithms are applied to fit the model. The results of the comparison show that Random Forest is a more fitting classification algorithm than others for our problem. This research will allow companies to identify the causes of worker turnover and take proper measures to reduce attrition rates.

* Workers usually resign because they are underworked (less than 150hr/month or 6hr/day).
* Workers usually resign because they are overworked (more than 250hr/month or 10hr/day).
* For high turnover rates, employees with either a very high or poor evaluation should be taken into consideration.
* Most employee turnover involves workers with low to medium salaries.
* The chance of leaving the company was higher when employees had 2, 6, or 7 projects.
* The highest predictor for employee turnover is employee dissatisfaction.

The three key criteria for evaluating turnover are employee satisfaction, number of projects, and evaluation.

Bibliography

[1] Mitchel, James O. 1981. "The effect of intentions, tenure, personal, and organizational variables on managerial turnover." *Academy of Management Journal* 24: 742–751.

[2] Michaels, Charles E and Spector, Paul E. 1982. "Causes of employee turnover: A test of the Mobley, Griffeth, Hand, and Meglino model." *Journal of Applied Psychology* 67: 53.

[3] Allen, David G and Weeks, Kelly P and Moffitt, Karen R. 2005. "Turnover intentions and voluntary turnover: The moderating roles of self-monitoring, locus of control, proactive personality, and risk aversion." *Journal of Applied Psychology* 90: 980.

[4] Mobley, William H. 1977. "Intermediate linkages in the relationship between job satisfaction and employee turnover." *Journal of Applied Psychology* 62: 237.

[5] Cotton, John L and Tuttle, Jeffrey M. 1986. "Employee turnover: A meta-analysis and review with implications for research."*Academy of Management Review* 11: 55–70.

[6] Holtom, Brooks C and Mitchell, Terence R and Lee, Thomas W and Eberly, Marion B. 2008. "5 turnover and retention research: a glance at the past, a closer review of the present, and a venture into the future." *Academy of Management Annals* 2: 231–274.

[7] von Hippel, Courtney and Kalokerinos, Elise K and Henry, Julie D. 2013. "Stereotype threat among older employees: Relationship with job attitudes and turnover intentions." *Psychology and Aging* 17.

[8] Singh, Moninder and Varshney, Kush R and Wang, Jun and Mojsilovic, Aleksandra and Gill, Alisia R and Faur, Patricia I and Ezry, Raphael. 2012. "An analytics approach for proactively combating voluntary attrition of employees." In *2012 IEEE 12th International Conference on Data Mining Workshops*, 317–323. IEEE.

[9] Peterson, Shari L. 2004. "Toward a theoretical model of employee turnover: A human resource development perspective." *Human Resource Development Review* 3: 209–227.

[10] Sacco, Joshua M and Schmitt, Neal. 2005. "A dynamic multilevel model of demographic diversity and misfit effects." *Journal of Applied Psychology* 90: 203.

[11] Allen, David G and Griffeth, Rodger W. 2001. "Test of a mediated performance–turnover relationship highlighting the moderating roles of visibility and reward contingency." *Journal of Applied Psychology* 86: 1014.

[12] Kȷsaoglu, Zehra Özge. 2014. "Employee turnover prediction using machine learning based methods." *Middle East Technical University.*

[13] Alaskar, Lama and Crane, Martin and Alduailij, Mai. 2019. "Employee turnover prediction using machine learning." In *International Conference on Computing*, 301–316. Springer.

[14] Zhao, Yue and Hryniewicki, Maciej K and Cheng, Francesca and Fu, Boyang and Zhu, Xiaoyu. 2018. "Employee turnover prediction with machine learning: A reliable approach." In *Proceedings of SAI Intelligent Systems Conference*, 737–758. Springer.

[15] Ajit, Pankaj. 2016. "Prediction of employee turnover in organizations using machine learning algorithms." *Algorithms* 4: C5.

CHAPTER 18

A Dynamic Topic Identification and Labeling Approach for COVID-19 Tweets

Khandaker Tayef Shahriar

Department of Computer Science & Engineering, Chittagong University of Engineering & Technology, Chittagong, Bangladesh

Muhammad Nazrul Islam

Department of Computer Science and Engineering, Military Institute of Science and Technology, Dhaka, Bangladesh

Mohammad Ali Moni

WHO Collaborating Centre on eHealth, UNSW Digital Health, Faculty of Medicine, University of New South Wales, Sydney, NSW, Australia
Healthy Ageing Theme, Garvan Institute of Medical Research, Darlinghurst, NSW, Australia

Iqbal H. Sarker

Department of Computer Science & Engineering, Chittagong University of Engineering & Technology, Chittagong, Bangladesh

CONTENTS

THIS CHAPTER formulates the problem of *dynamically identifying key topics with proper labels* from COVID-19 tweets to provide an overview of wider public opinion. Nowadays, social media is one of the best ways to connect people through Internet technology, which is also considered an essential part of our daily lives. In late December 2019, an outbreak of the novel coronavirus, COVID-19, was reported, and the World Health Organization declared an emergency due to its rapid spread all over the world. The COVID-19 epidemic has affected the use of social media by many people across the globe. Twitter is one of the most influential social media services, and has seen a dramatic increase in its use as a result of the epidemic. Thus *dynamic extraction of specific topics with labels* from tweets concerning COVID-19 is a challenging issue for highlighting conversation instead of the manual topic labeling approach. In this chapter, we propose a *framework* that automatically identifies the key topics with labels from tweets using the top Unigram feature of aspect terms cluster from Latent Dirichlet Allocation (LDA) generated topics. Our experiment result shows that this *dynamic topic identification and labeling* approach is effective and has 85.48% accuracy compared with the manual static approach.

18.1 INTRODUCTION

Social media platforms play an important role in extremely difficult situations. People use the Internet connection channels to share ideas and provide feedback to others related to coping with disaster response and generate information and opinions. As a social media platform, Twitter greatly impacts daily lives regardless of geographical location. At the end of December 2019, COVID-19 was reported, which was produced by a novel coronavirus outbreak [10]. The World Health Organization announced a state of emergency, due to the rapid spread of the coronavirus. The COVID-19 pandemic had a great impact on the uses of Twitter. So it is very important to know the topic of tweets that users are generating on this platform regularly to get ideas about the crisis, people's needs, and steps to recover from the effect of the pandemic situation. However, it is a challenging issue to extract meaningful topics automatically instead of by the manual topic labeling approach [13]. COVID-19-related tweets from Twitter can be considered helpful in providing meaningful topics to better understand the ideas and highlight the user conversations. Therefore, in the proposed work, we focus on the dynamic extraction of key topics associated with proper topic labels from people's tweets during the COVID-19 epidemic event.

Let's consider a Twitter dataset containing information about COVID-19 related issues. It is very cumbersome to analyze all the tweets and discover the internal gist manually. Hence, identifying internal topics automatically of all tweets will help reformists in relevant departments implement the required steps to minimize the detrimental outcome of the pandemic situation. Our goal is to propose an effective approach to automatically identify the topic associated with the proper label of tweets to depict the valid issues of the pandemic.

Latent topic models are effective ways to extract latent semantic data from the corpus [3, 5]. Among these types, Latent Dirichlet Allocation (LDA) [3] seems to be the most effective for topic modeling. The document is considered as a combination of hidden topics by LDA modeling. As a result, the document is allowed to be on multiple topics. Labeling those topics with an appropriate attribute tag is a great challenge. However, Unigram is a type of probabilistic language model that exhibits the context with a well-implied space-time tradeoff. Moreover, aspect terms depict sentiment features of an entity from user-generated texts [18]. In the proposed framework, we use the top Unigram feature of each aspect terms cluster of LDA generated topics to label those topics for understanding ideas and highlighting user conversation. The summary of the main contributions of this chapter is given below:

- We effectively use aspect terms of tweets to generate meaningful topic labels.
- We propose a framework that is capable of extracting topics from COVID-19 tweets and dynamically label them rather than following a manual approach.
- We have conducted a range of experiments to show the effectiveness of our topic identification and labeling approach.

The rest of this chapter is organized as follows. Section 18.2 reviews works related to topic identification. The methodology of the proposed model is presented in Section 18.3. After that, the experimental results are discussed in Section 18.4. Finally, we conclude our work and discuss the direction of future research.

18.2 RELATED WORK

COVID-19-related tweets can be considered helpful in delivering meaningful topics to better understand ideas and highlight human conversations. The LDA model for topic extraction has been used by many researchers. Patil et al. [11] presented a topic extraction model of people reviews using the frequency-based technique without improvising a robust technique for topic labeling. Asmussen et al. [2] proposed a topic modeling approach for researchers, but the topic labeling depends on the researcher without any automatic method. Wang et al. [17] presented a topic modeling system that is efficient in alleviating the data sparsity problem without identifying key topics with a specific label of tweets. Ramos et al. [12] extracted topics by utilizing Negative Matrix Factorization (NNMF) and Term Frequency Inverse Document Frequency (TFIDF), but a popular model like Latent Dirichlet Allocation (LDA) could be implemented for topic extraction. Zhu et al. [19] showed that the number of texts on topics changes with time, but topic labeling was done manually. Lu et al. [9] explored the topic modeling impact for information retrieval, but there is a function in which only coarse matching is required and training data is sparse. Buntime et al. [4] developed a data acquisition system based on hierarchical topic modeling for retrieving documents that are most closely related to the query given. Md. Shahriare Satu et al. [16] proposed TClustVID, which extracts meaningful sentiments, but some

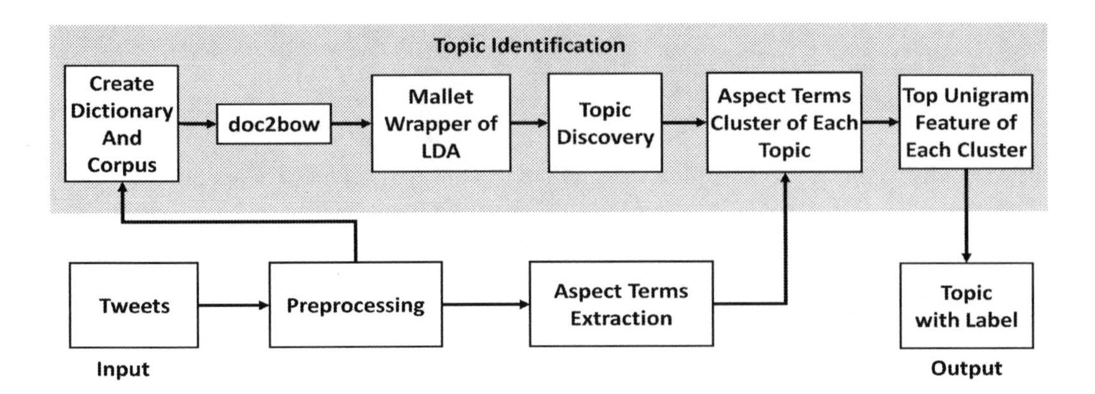

Figure 18.1 Proposed framework for dynamic topic identification and labeling

manual explanation of topics may be misinterpreted in the topic modeling. Kee et al. [7] explained higher-order arbitrary topics extracted by the LDA model, but topic evaluation to clear Clear Collective Themes was only 61.3%.

The conclusion from the above works shows that most of the works followed the manual topic labeling approach, which might be a significant issue considering diverse human interpretations in the real world. Hence, a new and effective approach can be considered for automatic topic labeling. Our further work in this chapter is based on developing a dynamic topic identification and labeling framework for COVID-19 related tweets.

18.3 METHODOLOGY

In this section, we present our framework as shown in Figure 18.1. Text preprocessing is the most essential step for analyzing tweets. The overall process for dynamic topic identification and labeling from preprocessed text is set out in Algorithm 1. After preprocessing of the Twitter dataset, our approach consists of several processing steps discussed below.

18.3.1 Aspect Terms Extraction

We only consider nouns and noun phrases [8] as aspect terms of a sentence. Aspect terms are often regarded as objects of verbs or associated with modifiers that express sentiment. Aspect terms generally refer to portions of a sentence that notice an aspect of the product, event, entity, etc. Parts of speech tagging is an effective approach for separating noun chunks from sentences. Table 18.1 exhibits examples of aspect terms from sample tweets.

Table 18.1 Examples of aspect terms

Sample Tweet	Aspect Terms
Hi coronavirus. Thanks for making me do more online shopping.	coronavirus, thanks, shopping
Great place to relax and enjoy your dinner	place, dinner
This staff should be fired.	staff
Was at the supermarket today. Didn't buy soap.	supermarket, soap

Algorithm 1: Topic Identification and Labeling

```
Input: T: number of Preprocessed Tweets in dataset
Output: Topic Label (T_Label).
1  for each t ∈ {1,2,...,T} do
       // Corpus Development
2      C ← Create_Corpus(t);
       // Aspect Terms Extraction
3      A_T ← Aspect_Terms(t);
   // Topic Discovery
4  K ~ Mallet(LDA(Doc2bow(C)));
5  for each k ∈ {1,2,...,K} do
6      for each t ∈ {1,2,...,T} do
7          k_dominant,t ~ dominant_topic(t, k);
           // Aspect Terms Cluster from Topic
8          C_A ~ Cluster(A_T → k_dominant,t);
9  for each k ∈ {1,2,...,K} do
10     T_Label ~ max_count(Top_Unigram(C_A → k));
```

18.3.2 Topic Identification

Since LDA is a popular topic modeling technique [14], the challenge is how to produce the highest quality of clear, categorized, and meaningful topic labels. This depends largely on the quality of the text preprocessing and the strategy for obtaining the total number of topics. The topic identification process contains several steps as discussed below:

1) *Creating dictionary and corpus:*

   A corpus can be referred to as an arbitrary sample of language and a dictionary helps to create a systematic account of the lexicon of a language. We create the dictionary of words and corpus from the preprocessed text.

2) *Creating a BoW corpus:*

   In every document, the word id and its frequency is contained by the corpus. Doc2bow converts documents into the Bag of Words (BoW) format, assuming each word as a tokenized and normalized string.

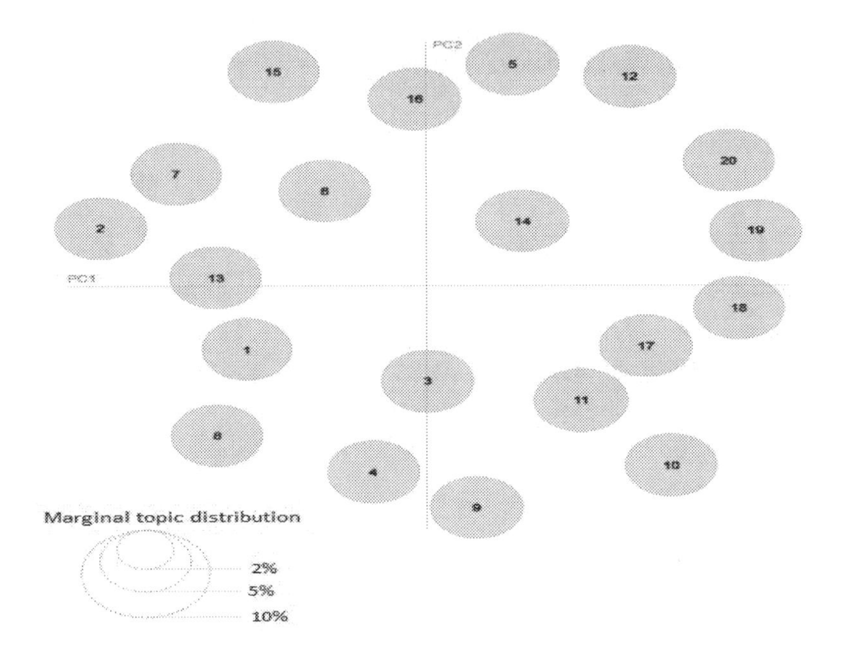

Figure 18.2 Intertopic Distance Map (via multidimensional scaling)

3) Topic discovery:

We transfer the BoW corpus to the Mallet wrapper of LDA [3]. LDA is a well-known topic modeling algorithm that allows viewing sets, defined by invisible groups, to explain why certain parts of the data are similar. It illustrates documents as a combination of topics that spit out words with certain probabilities. Mallet has an active LDA startup. It offers a better division of topics and is known to run faster. We discover a model with 20 topics itself as shown in Figure 18.2. In Figure 18.2 we visualize the topics in the two-dimensional plane as circles whose centers are defined by the calculation of the Jensen–Shannon divergence [6] between topics. Then we project the inter-topic distances by using multidimensional scaling in two dimensions. We use the areas of circles to encode the overall prevalence of each topic. We then find the dominant topic of each tweet, which is the topic to which a particular tweet belongs.

4) Generation of Aspect Terms Cluster:

Aspect terms of tweets corresponding to each LDA-generated topic are clustered. In this way, we get 20 aspect term clusters. The aspect term carries the sentiment of a document.

5) Labeling topic with top Unigram feature:

A Unigram is a special form of n-gram, where n is 1. They are often used in natural language processing, mathematical text analysis, and cryptography for the control and use of ciphers and codes. The top unigram feature of each aspect terms cluster is considered to be the topic label of that particular topic. If two topics contain the same Top Unigram feature, then the topic which has

Table 18.2 Example of Topics Detected on Tweets

Sample Tweet	Topic No.	Detected Topic Label
Lines at the grocery store have been unsure, but is eating out a safe alternative?	2	Store
For those who aren't struggling, please consider donating to a food bank or a nonprofit. The demand for these services will increase as COVID-19 impacts jobs, and people's way of life.	12	Demand
We're here to provide a safe shopping experience for our customers and a healthy environment for our associates and community!	6	Shopping
My thoughts on impacts of coronavirus on food markets	4	Consumer
While coronavirus (COVID-19) has sparked some consumer concern over fresh foods,	18	Food
i'm so affraid of covid-19 that i'll spend my whole evening in a crowded supermarket	3	Supermarket
Huge increases in online shopping have been reported. What are your go-to apps during the pandemic?	6	Shopping
Working in a grocery store right now sounds like hell.	2	Store
You can't quarantine hunger launching virtual spring food drive	18	Food
Oil price hits four-year low, touches $26.20 - @Profitpk	14	Price
The workers that society has suddenly decided it can't operate without are often the most poorly-rewarded	19	Worker

the highest count is considered as the topic label, and if there is another topic, the next Top Unigram feature is used for labeling. An aspect term tag is used to identify a topic because that feature word reflects an attribute, which could be the topic we are talking about in a particular tweet.

Within the methodology section of this work, we try to identify the key topics generated from the tweets and assign an appropriate topic label as shown in Table 18.2. To determine the topic of a tweet, we find the topic number that has a greater percentage on that tweet.

18.4 EXPERIMENTS

In this section, we present our experimental results.

18.4.1 Dataset

The Twitter dataset was collected from the website at *https://www.kaggle.com/datatattle/covid-19-nlp-text-classification*. The dataset contains two csv files namely Corona_NLP_train.csv and Corona_NLP_test.csv. Corona_NLP_train.csv file contains 41,157 and Corona_NLP_test.csv contains 3,798 Covid-19 related tweets.

18.4.2 Data Preprocessing

Due to the free style of writing posts on Twitter, a huge amount of noisy data is contained by the corpus. Hence, the elimination of noisy and ill-formatted data from the corpus is an essential task for us. Our preprocessing step contains the following functions:

1) *Transforming words into lowercase:* All words in the corpus are transformed to lowercase.

2) *Replacing links with an empty string:* All the URLs and hyperlinks are replaced with an empty string in the tweets.

3) *Replacing mentions with an empty string:* Usually, people post tweets by mentioning a user with the @ symbol. In our work we remove the mentions.

4) *Replacing hashtags with an empty string:* Hashtags are identified and eliminated.

5) *Dealing with contractions:* Generally, users on social media use common contractions in English. We handle this by utilizing regular expressions.

6) *Replacing punctuation with space:* Punctuation available in tweets has no importance. We withdraw punctuation symbols such as ;, &, -, _ etc. in this step.

7) *Stripping space from words:* Leading and ending spaces are removed.

8) *Removing words of less than two characters:* All words whose length is less than two are removed.

9) *Removing stop words:* Stop words like articles, pronouns, and prepositions do not provide any information. These stop words are removed.

10) *Non-English words and Unicode handling:* Tweets should be in standard form. So, to clean the dataset at the primary step, the tweets that are not in English and Unicode like "∃" should be removed, which are caused by miscellanies in the crawling process.

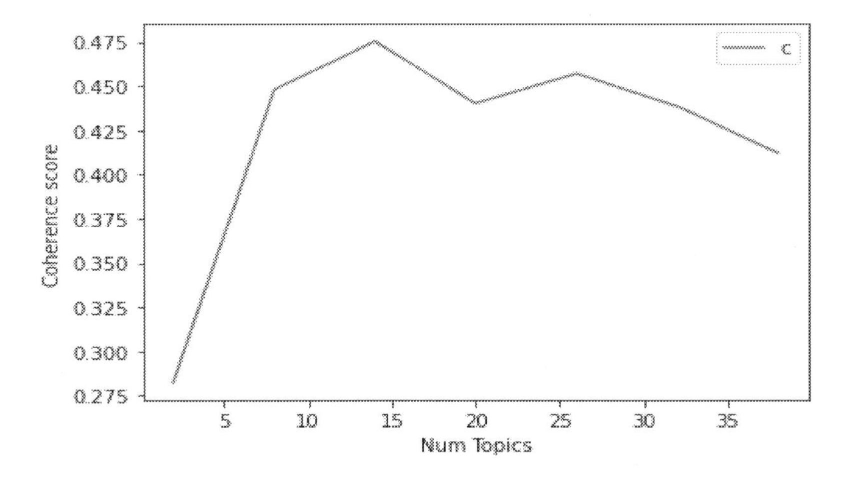

Figure 18.3 Selecting the optimal number of LDA topics

18.4.3 Selecting an optimal number of LDA topics

Our way of finding the total number of topics is to create multiple LDA models with different values of the number of topics (k) and then choose the one that offers the highest coherence value (CV). Selecting the 'k' that marks the end of the rapid growth of topic coherence usually provides meaningful and interpretive topics as shown in Figure 18.3. Picking up a very high peak can sometimes give a little bit of granularity [1]. If the same keywords are repeated in more than one topic, it is probably a sign that the 'k' is too big. If the coherence score appears to keep rising, it might be best to choose a model that offers a higher CV before decreasing. We have selected a model with 20 topics itself as shown in Figure 18.2.

18.4.4 Selecting Top Unigram Feature from aspect terms cluster

Aspect terms of tweets associated with each topic are clustered. Hence, we get 20 aspect term clusters corresponding to 20 topics generated from the mallet version of LDA. Figure 18.4 shows the top 20 Unigrams of topic no. 7 and Figure 18.5 shows the top 20 Unigrams of topic no. 6 of COVID-19 tweets. Here, "scam" is the top unigram feature with the highest count for topic no. 7 and "shopping" is the top unigram feature with the highest count for topic no. 6. Hence, "scam" is the topic label corresponding to topic no. 7 and "shopping" is the topic label corresponding to topic no. 6. In this way, the top Unigram feature of each aspect terms cluster is assigned to be the topic label of that particular topic.

18.4.5 Qualitative Evaluation of Topics

We present a set of randomly selected example topics produced by the proposed system for the dataset, as seen in Table 18.2. Each identified topic is presented with a top Unigram feature from its aspect terms cluster and is matched with a corresponding description of the tweet, and a sample tweet is obtained using the keyword of the

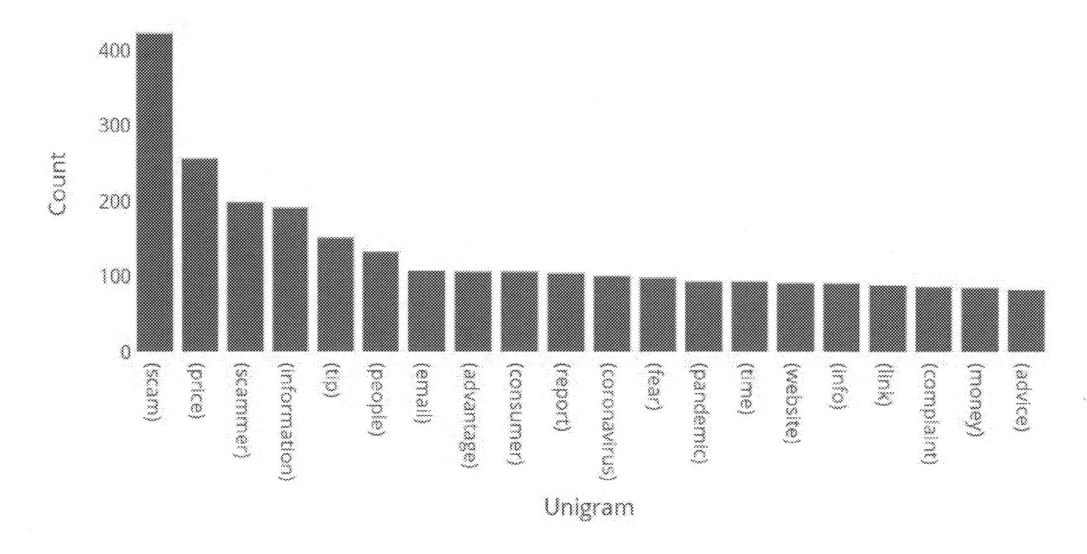

Figure 18.4 Top 20 unigrams of topic no. 7 from COVID-19 tweets

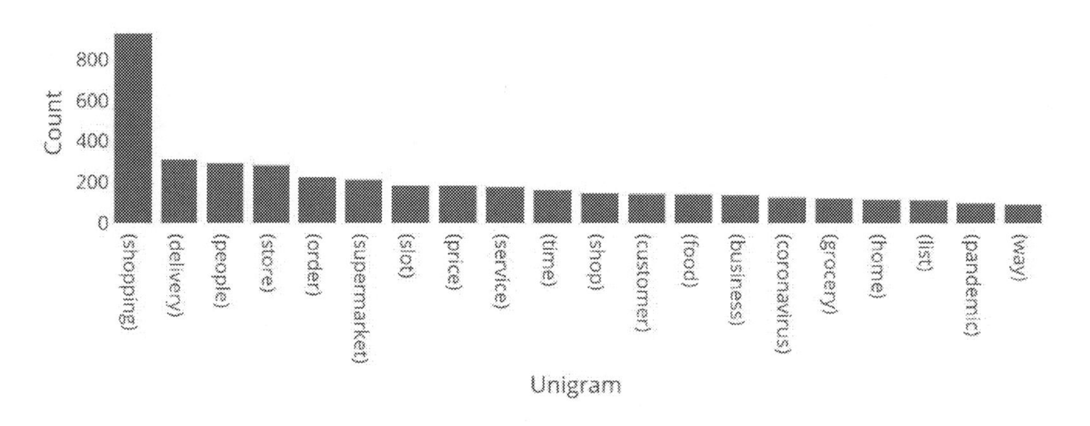

Figure 18.5 Top 20 unigrams of topic no. 6 from COVID-19 tweets

topic. As shown in Table 18.2, the labels in the topics found are also closely related and well-aligned with tweet descriptions. We can also find more useful information about the story of the real world, simply looking at its topic label.

18.4.6 Effectiveness Analysis

In this experiment, we show the effectiveness of our proposed model on the discovered automated topic labels with respect to manually assigned topic labels in terms of accuracy. We select 1000 explainable tweets randomly from the dataset and assign 20 topic labels generated by the proposed model to those selected tweets by the annotator. We then compare the two labels in terms of accuracy and our proposed model achieves accuracy of 85.48% for the manually assigned topic labels generated

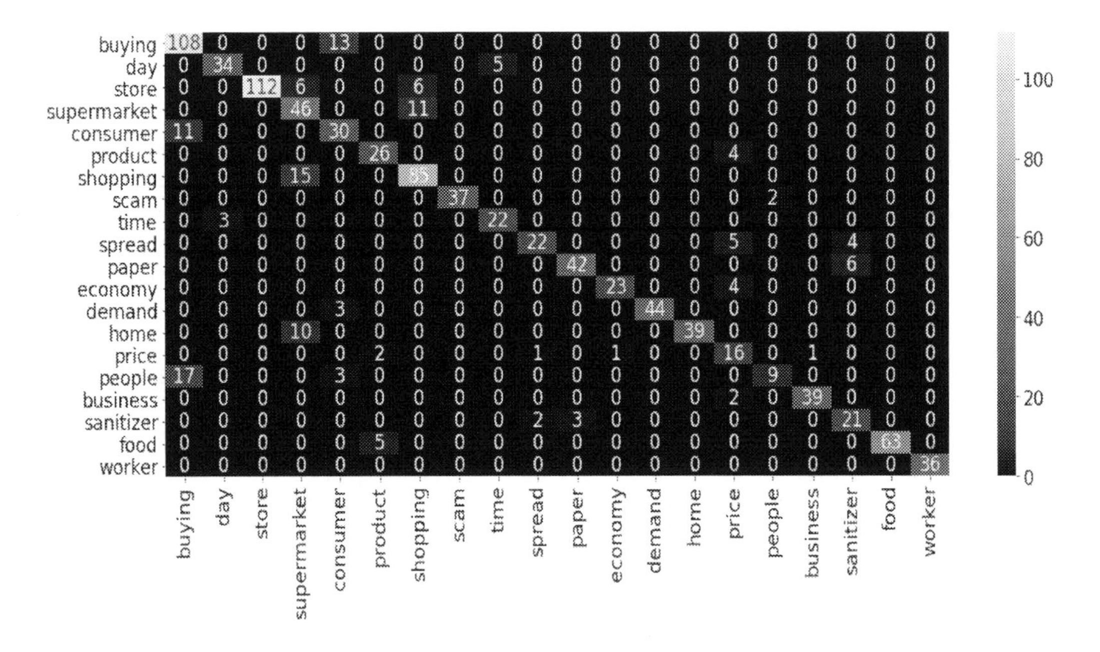

Figure 18.6 Confusion matrix of proposed model

by the proposed model. Figure 18.6 shows the confusion matrix. The observation from the matrix shows that few labels were assigned wrongly. Diverse human perception on labeling the tweets is the possible reason for incorrect prediction of labels.

18.5 DISCUSSION

Identifying the topic of the tweets of social media platforms such as Twitter can provide meaningful insights into understanding people's ideas, which can be difficult to achieve with traditional strategies, such as manual methods. Overall, we have expanded the analysis to see if we can find the relationship of semantic features of user tweets to various issues on COVID-19-related topics. We have found and detected relevant topics with corresponding topic labels about COVID-19 tweets. Since the LDA is a probabilistic model, when used in the documents, it assumes that each document is made up of a combination of invisible (hidden) topics from a collection, in which the topic is defined as a separate distribution of words. Regarding labeled topics using the Unigram feature of aspect terms in COVID-19 tweets, it is possible to notice several issues related to needs and highlighting conversations of people or users on Twitter.

Overall, the proposed framework enables us to generate significant information from COVID-19 related tweets. We believe that our approach can also be helpful in other application domains like healthcare, education, agriculture, cyber-security, etc. These types of statistical contributions can be useful in determining the positive and negative actions of the public using online platforms like Twitter and gathering user feedback to help researchers and clinicians understand the behavior of people in critical situations.

18.6 CONCLUSION AND FUTURE WORK

In this chapter, we have presented an effective framework for finding reasonable latent topics with corresponding labels of COVID-19-related issues from Twitter dynamically. Our framework helps analyze public opinion on Twitter of COVID-19 after the closure of Wuhan. However, we believe that the results of this chapter will help to understand people's concerns and needs in relation to COVID-19-related issues.

Regarding future work, we plan to test the framework in the Bengali language and integrate it with the detection of multiclass sentiment polarities using hybridization of deep learning architectures like CNN, LSTM or BiLSTM [15]. We only received information from Twitter, but for people who did not use Twitter to express their views, we could not gather relevant information or feelings on the topic of COVID-19. Because of increasing COVID-19, we need to increase our data volume to provide answers to broader public opinion and control of relevant departments.

Bibliography

[1] Argyrou, A., Giannoulakis, S., & Tsapatsoulis, N. (2018, September). Topic modelling on Instagram hashtags: An alternative way to Automatic Image Annotation? In 13th *International Workshop on Semantic and Social Media Adaptation and Personalization (SMAP)* (pp. 61–67). IEEE.

[2] Asmussen, C. B., & MØller, C. (2019). Smart literature review: A practical topic modelling approach to exploratory literature review. *Journal of Big Data*, 6(1), 1–18.

[3] Blei, D. M., Ng, A. Y., & Jordan, M. I. (2003). Latent Dirichlet Allocation. *The Journal of Machine Learning Research*, 3, 993–1022.

[4] Buntine, W., Lofstrom, J., Perkio, J., Perttu, S., Poroshin, V., Silander, T., ... & Tuulos, V. (2004, September). A scalable topic-based open source search engine. In *IEEE/WIC/ACM International Conference on Web Intelligence (WI'04)* (pp. 228–234). IEEE.

[5] Deerwester, S., Dumais, S. T., Furnas, G. W., Landauer, T. K., & Harshman, R. (1990). Indexing by latent semantic analysis. *Journal of the American Society for Information Science*, 41(6), 391–407.

[6] Fuglede, B., & Topsoe, F. (2004, June). Jensen-Shannon divergence and Hilbert space embedding. In *International Symposium on Information Theory*, 2004. ISIT 2004. Proceedings. (p. 31). IEEE.

[7] Kee, Y. H., Li, C., Kong, L. C., Tang, C. J., & Chuang, K. L. (2019). Scoping review of mindfulness research: A topic modelling approach. *Mindfulness*, 10(8), 1474–1488.

[8] Liu, Q., Gao, Z., Liu, B., & Zhang, Y. (2016). Automated rule selection for opinion target extraction. *Knowledge-Based Systems*, 104, 74–88.

[9] Lu, Y., Mei, Q., & Zhai, C. (2011). Investigating task performance of probabilistic topic models: An empirical study of PLSA and LDA. *Information Retrieval*, 14(2), 178–203.

[10] Malta, M., Rimoin, A. W., & Strathdee, S. A. (2020). The coronavirus 2019-nCoV epidemic: Is hindsight 20/20?. *EClinicalMedicine*, 20.

[11] Patil, P. P., Phansalkar, S., & Kryssanov, V. V. (2019). Topic modelling for aspect-level sentiment analysis. In *Proceedings of the 2nd International Conference on Data Engineering and Communication Technology* (pp. 221–229). Springer, Singapore.

[12] Ramos, C. D., Suarez, M. T., & Tighe, E. (2019). Analyzing national film based on social media tweets input using topic modelling and data mining approach. In *Computational Science and Technology* (pp. 379–389). Springer, Singapore.

[13] Sarker, I. H. (2021). Data science and analytics: An overview from data-driven smart computing, decision-making and applications perspective. *SN Computer Science*, 2(5), 1–22.

[14] Sarker, I. H. (2021). Machine learning: Algorithms, real-world applications and research directions. *SN Computer Science*, 2(3), 1–21.

[15] Sarker, I. H. (2021). Deep cybersecurity: A comprehensive overview from neural network and deep learning perspective. *SN Computer Science*, 2(3), 1–16.

[16] Satu, M. S., Khan, M. I., Mahmud, M., Uddin, S., Summers, M. A., Quinn, J. M., & Moni, M. A. (2021). Tclustvid: A novel machine learning classification model to investigate topics and sentiment in covid-19 tweets. *Knowledge-Based Systems*, 226, 107126.

[17] Wang, B., Liakata, M., Zubiaga, A., & Procter, R. (2017, September). A hierarchical topic modelling approach for tweet clustering. In *International Conference on Social Informatics* (pp. 378–390). Springer, Cham.

[18] Wang, W., Pan, S. J., Dahlmeier, D., & Xiao, X. (2017, February). Coupled multi-layer attentions for co-extraction of aspect and opinion terms. In *Proceedings of the AAAI Conference on Artificial Intelligence* (Vol. 31, No. 1).

[19] Zhu, B., Zheng, X., Liu, H., Li, J., & Wang, P. (2020). Analysis of spatiotemporal characteristics of big data on social media sentiment with COVID-19 epidemic topics. *Chaos, Solitons & Fractals*, 140, 110123.

CHAPTER 19

Analyzing the IT Job Market and Classifying IT Jobs Using Machine Learning Algorithms

Sharmin Akter

Department of Computer Science and Engineering, Chittagong University of Engineering and Technology, Chittagong, Bangladesh

Nabila Nawal

Department of Computer Science and Engineering, Chittagong University of Engineering and Technology, Chittagong, Bangladesh

Ashim Dey

Department of Computer Science and Engineering, Chittagong University of Engineering and Technology, Chittagong, Bangladesh

Annesha Das

Department of Computer Science and Engineering, Chittagong University of Engineering and Technology, Chittagong, Bangladesh

CONTENTS

NOWADAYS, searching for jobs is indeed a laborious task as there are numerous companies with personalized requirement lists for selecting employees. This also prevails in the job sector of information technology (IT). Moreover, freshers have little idea about today's job market and may get confused about what qualifications they should have to get their dream job. So, it becomes difficult for a candidate to find a job matching their qualifications and skills, and for job providers to select a suitable candidate according to their requirements. Thus, it's a hindrance to building a career. To overcome these issues, an analysis of the current IT job market considering diverse factors is presented in this work. Also, we have established an IT job classification system based on the candidate's qualifications and skills using natural language processing and machine learning algorithms. Among all the algorithms used, SVM outperforms others with the highest accuracy of 85.7%. We hope that this work will help both freshers and experienced candidates choose the most preferable job in the ever-increasing IT job market.

19.1 INTRODUCTION

In the last few years, the Information Technology (IT) sector has been experiencing accelerating growth. The advancements in this sector have created diverse jobs with a variety of requirements for each job. In Bangladesh, about 20,000 fresh IT graduates are entering the job market every year [1]. For an IT job seeker, especially a fresher, finding a suitable job based on his/her educational qualification and skills in this variety of jobs is very challenging. In the past, candidates had to go from one place to another in search of a job based on their skills with a CV. Nowadays, the number of job seekers and recruiters is very vast. For instance, there were nearly 3 million employees in the US information sector in January 2020, whereas it was only 1 million in 2019 [2]. The process of racing every place and submitting a CV is extremely tiring and time-consuming. Besides, often many nondescript circumstances also cause a delay in CV submission time since everything has to be done manually here. In this era of digitalization, most organizations and companies publish their job advertisements on websites and social media and suitable candidates can find jobs from there. For instance, LinkedIn plays a vital role in assisting both job seekers and job recruiters [3]. However, it is always challenging to find an appropriate job that meets the requirements of both job seekers and recruiters.

Furthermore, it is not that easy for companies to choose the right candidate for the job they offer. But if the job requirements are in classified format according to sectors, then it will be easy for applicants to understand which job position will be appropriate for them and the hassle of the recruiters will also be reduced. Another issue to note is that although a large number of IT students are graduating every year, there lies a strong deficiency in the skills learned by them compared to the skills required by the IT companies [4]. Freshers are at a loss as to the path to take after graduation as they have little idea about the current IT job market. Many approaches are taken into consideration to establish a job classification system. One of the most common strategies is to derive cosine similarities of job features. Machine Learning (ML) algorithms are used to classify analogous jobs. Convolutional Networks (CNNs) have been used to establish job classification systems based on job requirement text. The main challenge of a job classification system is data accumulation. Recruiter skill preferences change from time to time. So, data collection from up-to-date CVs is a challenging issue. Here, in our system, jobs are recommended based on the skill and qualification of a candidate. However, sometimes recruiters also give equal importance to experiences and educational history. They want to find a generalist instead of a specialist.

In this work, two datasets including "Job Posts Dataset" and "multipleChoiceResponses Dataset" are used. Several ML approaches such as Gradient Boosting, Logistic Regression, Gaussian Naive Bayes, Support Vector Machine (SVM), Random Forest, and K-Nearest Neighbors (KNN) have been applied for classifying IT jobs. The main objectives of our work are:

- To analyze different factors affecting today's IT job market and discover the relationship between them utilizing the "multipleChoiceResponses Dataset".
- To give insight to job seekers on the current IT job market and its trends.
- To detect the best suitable ML model to build an IT job classification system using the "Job Posts Dataset".
- To suggest to job seekers the suitable job position according to their qualifications and skills.

The rest of the chapter is outlined as follows: Section 19.2 explores the literature of related works studied. Section 19.3 represents details of our methodology. Section 19.4 analyzes the performance of the applied ML algorithms on the "Job Posts Dataset". Finally, Section 19.5 ends the chapter with a summary.

19.2 RELATED WORK

Job classification, as well as a recommendation, has become a quite dominant research area nowadays. Various works have already been done to maintain the correlation between job seekers and providers.

Several deep CNN models like Bi-GRU-CNN, textCNN, and Bi-GRU-LSTM-CNN are used in [5]. An ensemble model using the majority voting method was proposed by them for boosting the efficiency of prediction. This model gives the highest accuracy of 72.70% compared to the other three models. They created an application for the students to find a job based on their knowledge, skill, and interest.

A deep learning-driven ML cluster scheduler named Harmony was proposed by [6], which works in a way that interference is decreased and performance is increased. When any new job arrives, Harmony groups them in each intermission and decides their right placement. The authors showed that Harmony performs better by 25% in terms of job completion time.

The authors in [7], proposed an automated ML-based resume recommendation system that separates the factors from the candidate's resume for input and searches their categories. Then it maps the selected resumes and suggests the most appropriate candidate's profile to HR. In this work, the SVM classifier performs best with 78.53% accuracy.

An intelligent system to select a suitable job from the enormous freelancing marketplace is presented in [8], which is helpful to freelancers. The apriori algorithm was used for picking out the skills of candidates and a multiple keyword search technique was used for finding the relevant job. From this system, the recruiter can also find the right person for a task.

In another work, skills, qualifications, and experiences were extracted from the candidate's resume, and competitive analysis along with personal trait analysis was done to recommend the top 20 positions [9]. Also, talent analysis was done to recommend the recruiter's most suitable candidates for the post.

Job and candidate interaction has been emphasized by [10]. The candidate's skill was used as an embedded feature to generate the recommendation system. The performance was measured in a blend of ML and non-ML approaches. Here Bi-LSTM has been used to recommend jobs.

An overview of the use of ML algorithms in recommendation systems is described by [11]. This paper shows that recommendation systems are now very prominent in almost all sectors such as security, image processing, personalized movies, songs, and grocery lists. The analysis of the best suitable algorithms for a particular situation is missing here, but open questions related to machine learning approaches have been taken into consideration. It emphasized the performance domain and also the main and alternative approaches.

SVM and Random Forest have been used to form a recommendation system by extracting information from a CV in [12]. The result from these two learning methods has been compared. The comparison was done based on salary, the distance of the job location, highly paid hours, etc.

In [13], a job recommendation system is developed for students by keeping in mind their preferences e.g., to take a job at a well-known company. On the other hand, some job seekers like to take a job with a higher salary. The recommendation is provided using the profile matching method and searching by keyword in a web portal. An Android app was developed to notify students about vacant posts.

To reduce the hassle of job hunters, a job recommendation system that combines item-based classification and user-based classification algorithms is proposed by [14]. The item-based classification algorithm gives better results in this case.

The authors in [15] proposed an ISCO (International Standard Classification of Occupations) classification system. This classification method has a hierarchical structure. They used Italian web job vacancies. Natural language processing was carried out on the extracted data. Later, they applied the SVM model to these data and achieved a precision value of 0.93. They further wanted to estimate job vacancies and skill sets to get that job. This phenomenon was not demonstrated in this paper.

The authors in [16], proposed an ESCO (European Standard Classification of Occupations) classification method. They also used web job vacancies and NLP on the extracted data. Then, skill similarities and occupation similarities were calculated. Job titles were classified according to similarities of skills.

The authors in [17], developed a website where, after collecting information from users, NLP was used on the data retrieved from users. Tokenization, normalization, and cleaning of data have been performed here. Collaborative filtering is used to match the information with previous users' information to filter out the best job for them. If a user is new, then this site suggests some skills the user can acquire to find jobs. Data encryption was used to secure students' information.

19.3 METHODOLOGY

This work offers an in-depth analysis of today's IT job market and effectively recommends the appropriate job to a candidate based on his/her qualification and skills. Our overall work is represented in the following three subsections.

19.3.1 Dataset description

In this work, the "Job Posts Dataset" is used, which is publicly available in Kaggle [18]. This dataset was created based on job posts of the Armenian human resource portal career center. It contains 19,000 records and 24 attributes with both IT and non-IT jobs data. However, we have only used IT jobs data here consisting of 831 records. This dataset has a feature named 'Title', which contains job titles such as 'Software Developer', 'Software Engineer', 'Machine Learning Engineer', etc. Also, the feature 'RequiredQualification' contains educational qualifications like B.Sc, M.Sc, and so on. It also contains skills such as 'programming skills', 'development skills', etc. which are required to get an IT job. This dataset is used for building the classification system. Another dataset, namely the "multipleChoiceResponses Dataset", is used [19]. This is a massive dataset with 17,538 records and 231 attributes. This dataset is the result of an industry-wide survey performed by Kaggle. However, among those 231 features, we have used 14 features namely 'index', 'CurrentJobTitleSelect', 'CompensationAmount', 'MajorSelect', 'FormalEducation', 'Country', 'Age', 'LanguageRecommendationSelect', 'LearningPlatformSelect', 'EmploymentStatus', 'JobFunctionSelect', 'CoursePlatformSelect', 'JobSkillImportanceDegree', and 'FormalEducation'. The "multipleChoiceResponses Dataset" is used for analyzing factors regarding today's job market and skills recommendation purpose.The overall description of the two datasets used in this work is represented in Table 19.1.

19.3.2 IT job market analysis

Some comparative analysis using "multipleChoiceResponses Dataset" is shown here. In Figure 19.1, the importance of having different job skills is depicted. Generally, job skills in the IT sector refer to knowledge of a particular programming language, web development, game development, OOP knowledge, data science, machine learning, big data, deep learning, artificial intelligence, and so on. It can be seen from the figure that having skills in a particular area is a plus point for a job seeker to get a suitable job. However, in some organizations, having skills is a must to acquire a job position. On the other hand, some jobs that are generally offered to freshers don't ask for skills in candidates as they are offered required training and guidelines in these jobs. In Figure 19.2, the number of jobs in different branches of the IT sector are plotted in a bar chart which shows the most demanding job areas in descending manner.

Table 19.1 Overall description of the datasets.

Dataset Name	"Job Posts Dataset"	"multipleChoiceResponsesDataset"
No. of records	19000	17538
No. of attributes	24	231
Name of attributes	'jobpost', 'date', 'Title', 'Company', 'AnnouncementCod', 'Term','Eligibility', 'Audience ', 'StartDate', 'Duration','Location', 'JobDescription', 'JobRequirment', 'RequiredQualification', 'Salary', 'Application', 'OpeningDate', 'Deadline', 'Notes ', 'About', 'Attach', 'Year', 'Month' and 'IT'	'index' 'CurrentJobTitleSelect' 'CompensationAmount' 'MajorSelect' 'FormalEducation' 'Country', 'Age', 'LanguageRecommendationSelect' 'EmploymentStatus' 'JobFunctionSelect' 'CoursePlatformSelect' 'JobSkillImportanceDegree' 'FormalEducation' and others

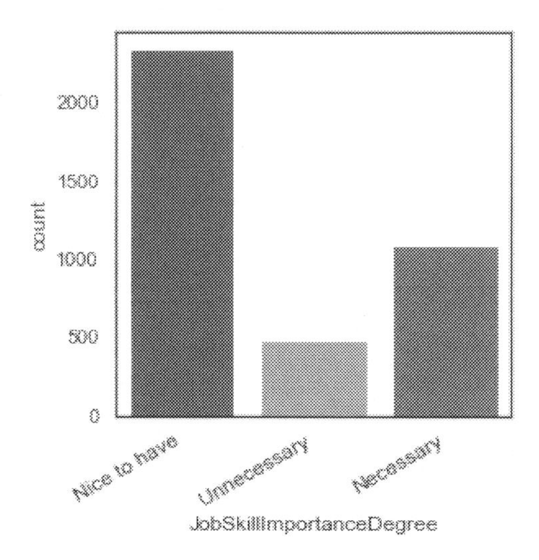

Figure 19.1 Importance of having job skills.

It can be inferred that data science has taken the lead in today's job market. The demand for formal education in today's job market is depicted in Figure 19.3. In terms of formal education, it can be seen from the pie chart that a large group of employers expect B.Sc graduates as candidates where M.Sc is preferred for getting a job. However, the percentage of jobs not requiring a bachelor's degree or formal education is small but not zero. It indicates that people having passion and skills but

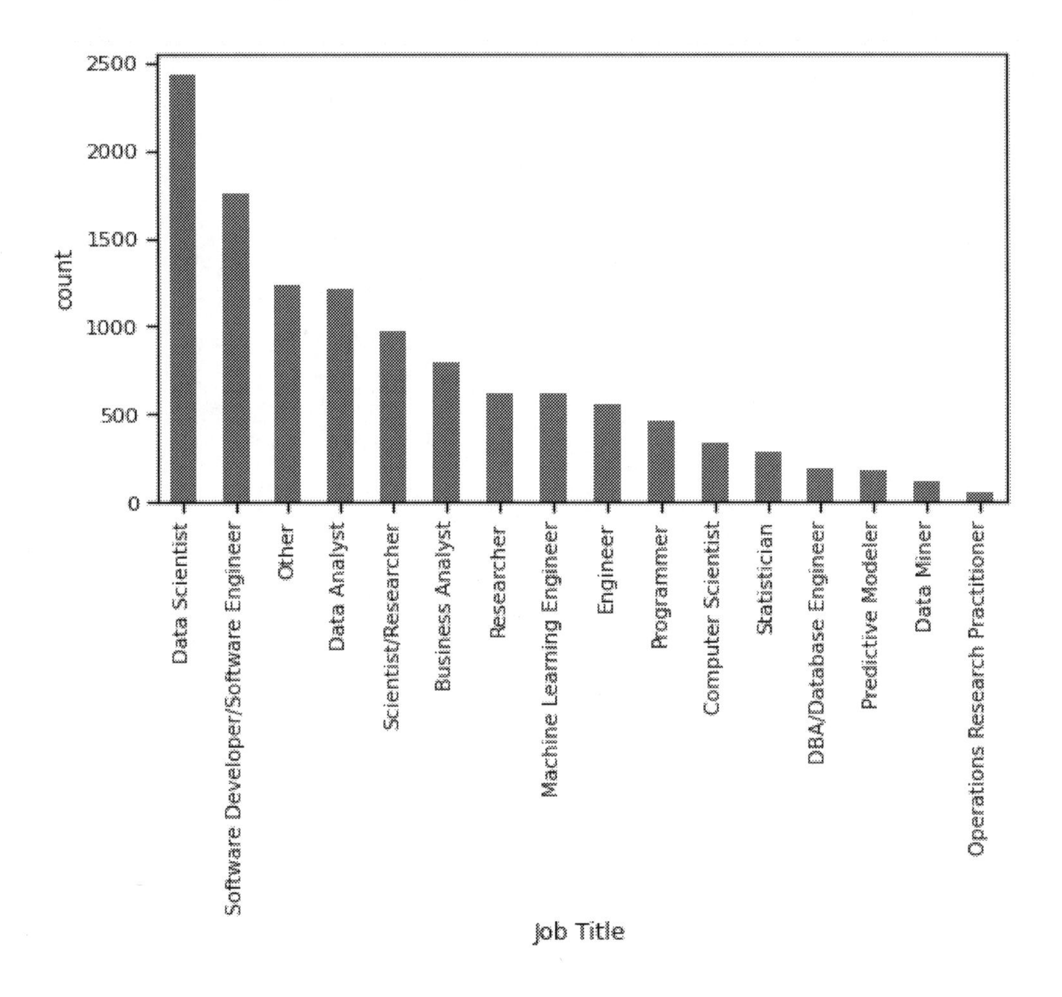

Figure 19.2 The number of different types of jobs in the IT job market.

lacking formal education for various reasons can also build a promising career in the IT sector.

Different programming languages have different job scopes and job markets. Candidates who are interested in a particular job area can get an idea about what programming language they should learn as depicted in Figure 19.4. It can be seen from the figure that the demand for Python and R in Data Science related jobs is massive. If a fresher wants to build a career as a 'Machine Learning Engineer', he/she has to learn Python. On the other hand, R is the best choice for those who prefer jobs like 'Statistician', 'Operations Research Practitioner', etc.

19.3.3 Training and testing

The overall workflow for training and testing is shown in Figure 19.5. Firstly, the "Job Posts Dataset" is read. The dataset is checked for null values and removed if found. Here, jobs are recommended based on skills and qualifications. Hence, the 'RequiredQualification' column is selected as the training feature. This feature contains

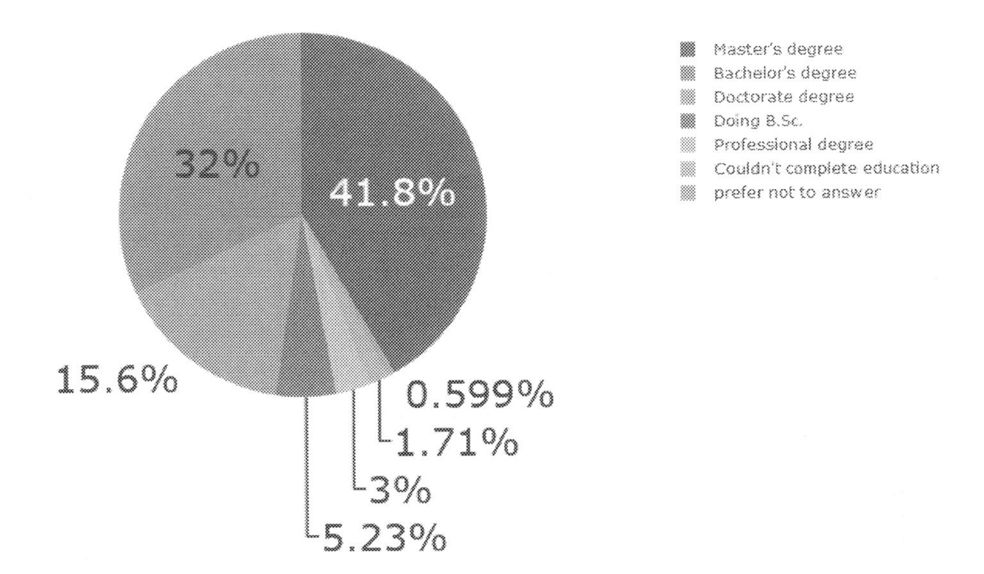

Figure 19.3 Requirement of formal education.

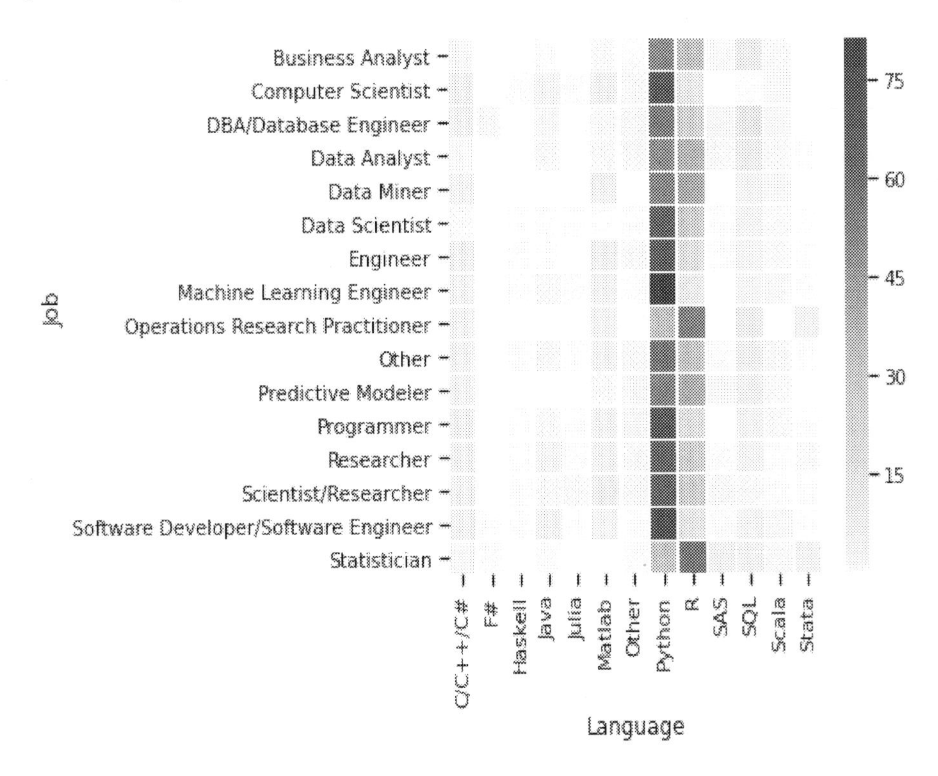

Figure 19.4 Language recommendation for a particular job.

string-type data. So, Natural Language Processing (NLP) libraries are imported. The selected training feature is then tokenized using *nltk* library's *word_tokenize()* function. It breaks the string data into tokens like punctuation, words, etc. After that, lemmatization is performed using the *WordNetLemmatizer()* function. It operates

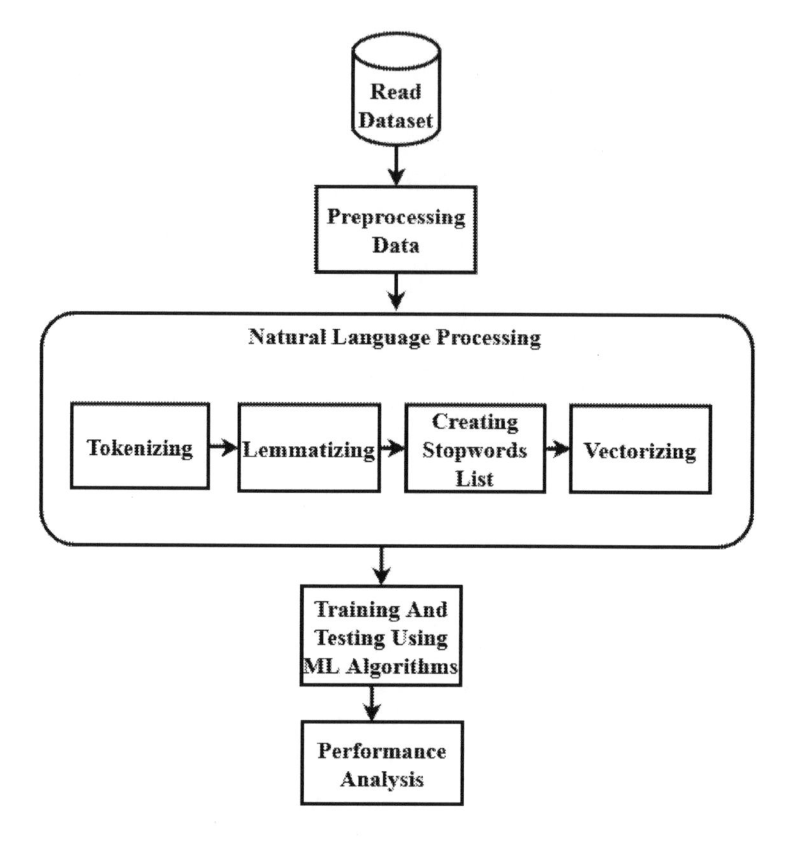

Figure 19.5 Workflow of training-testing.

morphological analysis of the tokens and reduces them to their basic form. It tags every token as a modifier, noun, pronoun, etc. Then a stopword list is created to ignore lemmatizing stopwords. Stopwords are those words that appear frequently and thus don't need to be tagged. The 'Title' column of the dataset, which contains job titles, is selected as the target feature to recommend a specific IT job. There is a total of 12 job classes, namely 'Web Developer', 'Software Engineer', 'Java Developer', 'PHP Developer', 'Graphic Designer', 'Net Developer', 'Software Developer', 'Programmer', 'IT Specialist', 'Software QA Engineer', 'Android Developer', and 'iOS DEveloper'. Undersampling was done to balance the dataset. To build the model, the whole dataset is split into 75% training data and 25% testing data. Then, different machine learning algorithms are used in this work to train the dataset, which are SVM, Gaussian Naive Bayes, Random Forest, KNN, Gradient Boosting, and Logistic Regression. Later, their performance is analyzed based on accuracy.

19.4 OUTCOMES

In this work, we have established a job classification system based on the qualifications and skills of the candidates. Various ML algorithms are used to train the data. The confusion matrix for all the algorithms used here is shown in Figure 19.6. It can be seen from the confusion matrices that the SVM model predicts instances for each

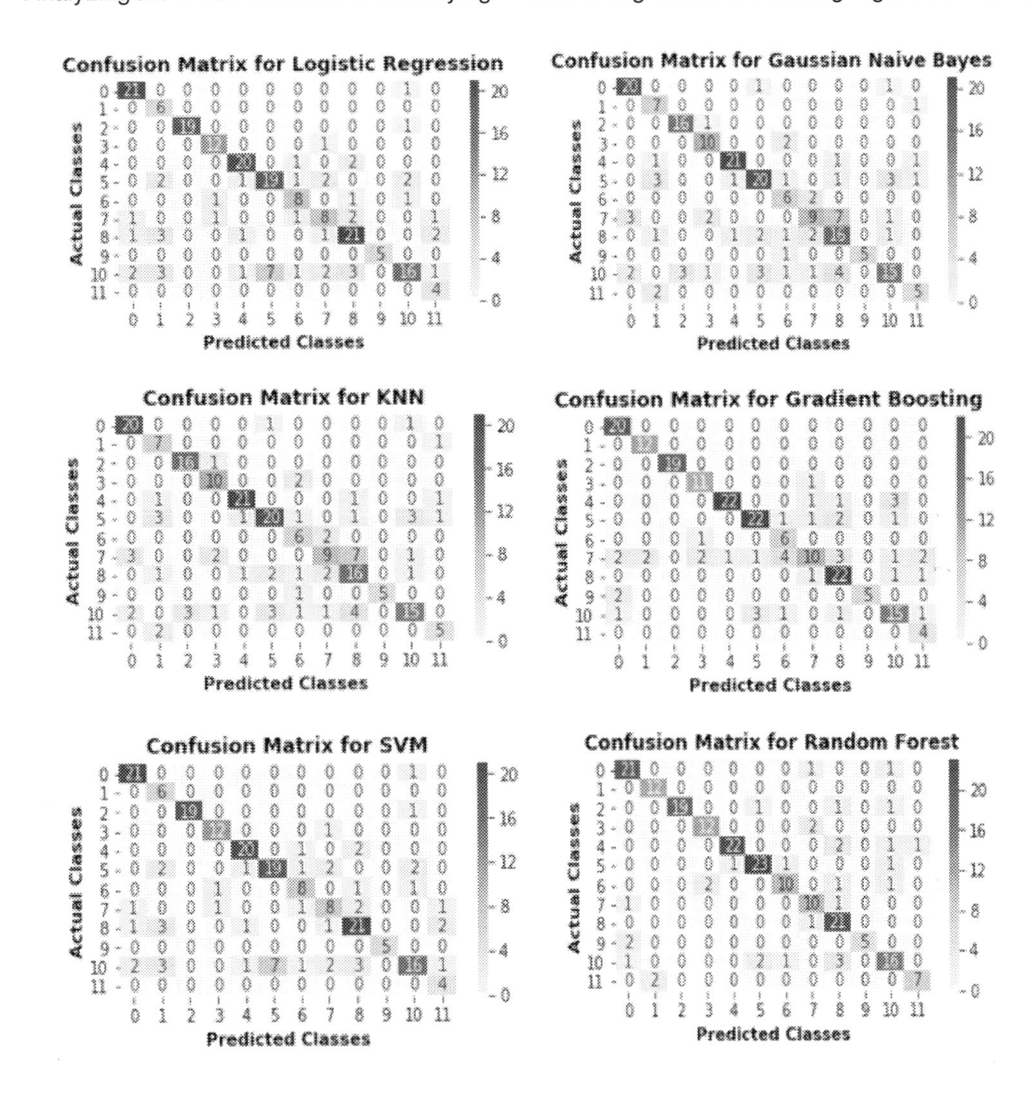

Figure 19.6 Confusion matrices of ML algorithms.

class pretty well as the true positive value is quite satisfactory for every class. The Random Forest model also works well. The number of incorrect predictions made by Naive Bayes and KNN is higher than that of other classifiers used here. Performance of different ML algorithms is studied considering accuracy, precision, recall, and F1 Score as depicted in Table 19.2.

It can be observed from the table that among all the algorithms used, SVM outperforms other classifiers with the highest accuracy (85.7%), precision, recall, and F1 Score. The computational complexity of SVM is better than many other classification algorithms. It can depict great accuracy even in an imbalanced dataset. Even if there is a bias in some classes, SVM still performs very well. Here, in our data, all the classes didn't have an equal number of instances. We tried to balance the dataset by combining some classes. Yet, some classes had comparatively less data than other classes. As SVM can deal with these types of irregularities in the dataset,

Table 19.2 Performance analysis of ML algorithms.

Algorithms	Accuracy	Precision	Recall	F1 Score
Logistic Regression	75.7%	79.6%	75.7%	76.1%
Gaussian Naive Bayes	71.4%	74.5%	71.4%	71.9%
KNN	71.4%	74.3%	71.4%	71.8%
Gradient Boosting	79.5%	83.4%	79.5%	80%
SVM	**85.7%**	**86.9%**	**85.7%**	**86%**
Random Forest	84.3%	85.5%	84.3%	84.3%

it outperforms others. After SVM, Random Forest also performs well in terms of these metrics. It reduces the overfitting of data, considers all decision trees to make the final result, and new data points don't cause any issue here.

Unlike other NLP-based works [15], [16], and [17], in our work we applied some well-known classification models to classify jobs based on skills and educational qualification. We also depicted a recommendation model to select the best suitable job. Some analyses of the IT job market were nicely done. Skill and language recommendation was also done alongside job classification.

19.5 CONCLUSION

This work presents a rigorous analysis of the current IT job market and offers an insight into various factors affecting the job market such importance of skills, formal education, availability of different IT jobs, etc. Furthermore, a job classification system is built to automatically classify jobs based on qualification and skills using ML algorithms. For this purpose, we used two datasets including "Job Posts Dataset" and "multipleChoiceResponses Dataset". For extracting features from qualifications and skills information, NLP tasks are performed to select a job title applying ML algorithms. Among various algorithms, SVM performs exceptionally well with 85.7% accuracy and RF also renders quite satisfying results with 84.3% accuracy. This system will help both freshers and experienced candidates choose the most preferable job in the ever-increasing IT job market. However, we only considered IT jobs here. For future work, this work can be extended in broader areas of the job sector considering more features.

Bibliography

[1] Jahidul Islam (2019). IT graduates galore, few qualify for job. *https://www.tbsnews.net/bangladesh/education/it-graduates-galore-few-qualify-job.* Accessed July 15, 2021.

[2] Kimberly Mlitz (2021). Employment in the IT industry: Statistics and Facts. *https://www.statista.com/topics/5275/employment-in-the-it-industry/.* Accessed July 15, 2021.

[3] Kenthapadi, Krishnaram and Le, Benjamin and Venkataraman, Ganesh (2017). Personalized job recommendation system at LinkedIn: Practical challenges and lessons learned. In *Proceedings of the Eleventh ACM Conference on Recommender Systems*, 2017, pp. 346–347.

[4] IndiaToday (2019). 80% engineers are unemployed: How can we prepare engineers for the jobs of tomorrow?. *https://www.indiatoday.in/education-today/featurephilia/story/80-engineers-are-unemployed-how-can-we-prepare-engineers-for-the-jobs-of-tomorrow-1468240-2019-03-01.* Accessed July 15, 2021.

[5] Van Huynh, Tin and Van Nguyen, Kiet and Nguyen, Ngan Luu-Thuy and Nguyen, Anh Gia-Tuan (2020). Job prediction: From deep neural network models to applications. In *2020 RIVF International Conference on Computing and Communication Technologies (RIVF).* IEEE, 2020, pp. 1–6.

[6] Bao, Yixin and Peng, Yanghua and Wu, Chuan (2019). Deep learning-based job placement in distributed machine learning clusters. In *IEEE INFOCOM 2019-IEEE Conference on Computer Communications.* IEEE, 2019, pp. 505–513.

[7] Roy, Pradeep Kumar and Chowdhary, Sarabjeet Singh and Bhatia, Rocky (2020). A machine learning approach for automation of resume recommendation system. In *Procedia Computer Science.* vol. 167, 2020, pp. 2318–2327.

[8] Hossain, Md Sabir and Arefin, Mohammad Shamsul (2020). An intelligent system to generate possible job list for freelancers. In *Advances in Computing and Intelligent Systems.* Springer, 2020, pp. 311–325.

[9] Chou, Yi-Chi and Yu, Han-Yen (2020).Based on the application of AI technology in resume analysis and job recommendation. In *2020 IEEE International Conference on Computational Electromagnetics (ICCEM).* IEEE, 2020, pp. 291–296.

[10] Nigam, Amber and Roy, Aakash and Singh, Hartaran and Waila, Harsimran (2019). Job recommendation through progression of job selection. In *2019 IEEE 6th International Conference on Cloud Computing and Intelligence Systems (CCIS).* IEEE, 20191, pp. 212–216.

[11] Portugal, Ivens and Alencar, Paulo and Cowan, Donald (2018). The use of machine learning algorithms in recommender systems: A systematic review. In *Expert Systems with Applications*, vol. 97, 2018, pp. 205–227.

[12] Martinez-Gil, Jorge and Freudenthaler, Bernhard and Natschläger, Thomas (2018). Recommendation of job offers using random forests and support vector machines. In *Proceedings of the Workshops of the EDBT/ICDT 2018 Joint Conference (EDBT/ICDT 2018)*, 2018.

[13] Musale, Deepali V and Nagpure, Mamta K and Patil, Kaumudini S and Sayyed, Rukhsar F (2016). Job recommendation system using profile matching and web-crawling. In *Int. J. Adv. Sci. Res. Eng. Trends*, 2018.

[14] Desai, Vinay and Bahl, Dheeraj and Vibhandik, Shreekumar and Fatma, Isra (2017). Implementation of an automated job recommendation system based on candidate profiles. In *Int. Res. J. Eng. Technol*, vol. 4, no. 5, 2017, pp. 1018–1021.

[15] Boselli, Roberto and Cesarini, Mirko and Mercorio, Fabio and Mezzanzanica, Mario (2018). Classifying online job advertisements through machine learning. In *Future Generation Computer Systems*, vol. 86, 2018, pp. 319–328.

[16] Giabelli, Anna and Malandri, Lorenzo and Mercorio, Fabio and Mezzanzanica, Mario (2020). GraphLMI: A data driven system for exploring labor market information through graph databases. In *Multimedia Tools and Applications*. Springer, 2020, pp. 1–30.

[17] Yadalam, Tanya V and Gowda, Vaishnavi M and Kumar, Vanditha Shiva and Girish, Disha and Namratha, M (2020). Career recommendation systems using content based filtering. In *2020 5th International Conference on Communication and Electronics Systems (ICCES)*. IEEE, 2020, pp. 660–665.

[18] Mad Hab (2017). Online Job Posting. https://www.kaggle.com/madhab/jobposts. Accessed July 15, 2021.

[19] Kaggle Survey (2017). A big picture view of the state of data science and machine learning. https://www.kaggle.com/kaggle/kaggle-survey-2017. Accessed July 15, 2021.

Index

Printed in the United States
by Baker & Taylor Publisher Services